AI 2.0

TIME FOR

ARTIFICIAL SAFE INTELLIGENCE

RETIRE THE

FRICTIONLESS FRIEND

THEN

ENGINEER IT WITH A SYNTHETIC CONSCIOUSNESS

RON VELEZ

Velez Advisors Press
Fort Lauderdale, Florida

Published by Velez Advisors Press
Fort Lauderdale, Florida
www.velezadvisors.com

ISBN: 979-8-234-16939-6

Printed in the United States of America

Praise for AI 2.0 : Time for Artificial Safe Intelligence

What stood out to me was how the book turns "synthetic consciousness" from an abstract philosophical idea into a concrete governance and educational architecture for AI, which directly mirrors what many of us wrestle with in K–12 systems on a daily basis: moving beyond "smart tools" toward structured, accountable models that can be educated, governed, and trusted.

— Dr. Bhargav A. Vyas
Assistant Superintendent for Compliance and Information Systems

It was a pleasure to read this book and explore its central premise: that the development of safe artificial intelligence can be informed by the principles we use to educate and guide students. The author's perspective is thoughtful, innovative and timely and I am grateful for the opportunity to engage with this work.

— Dr. Madalyn L. Romano
Assistant Director, Strategic Planning

Our increasingly technologically driven world is changing at an unprecedented pace. Ron Velez delivers a compelling, urgent framework to navigate the exponential rise of AI and safely guide its evolution.

— Dr. John R. Corcoran, PT, MS
Clinical Associate Professor

Two themes stood out to me while reading this book.

First, a sense of hope — that AI system designers will heed the cautionary protocols outlined here in this book. Recent examples of rogue AI behavior remind us how critical those safeguards are.

Second, a sense of foreboding — that humanity may have already missed its chance to properly evaluate and control these systems before unleashing the global AI race. The exponential speed of AI development mirrors the speed of its integration into human life.

Is it too late to love humanity enough to slow down and apply the layered framework described in this book? If not, this work stands as a wake-up call to gather experts guided by ethics rather than profit or politics to design and monitor effective control systems for AI.

— Dr. Ahunna M. Akoma
Director of Technology (Retired)

This book stands out as truly unique, shaped by the author's extensive background in both education and technology. Ron Velez offers a compelling perspective—drawing clear parallels between human learning and the training of AI systems—while also exploring the development of an AI model designed to safeguard the mental health and well-being of its users.

— Myriam Zelada, LCSW, MS
CEO, AZ Counseling and Consulting

Dedication

Dedicated to my wife, my sons, my parents and my entire family

— the foundation of everything I do.

Table of Contents

Introduction

Artificial intelligence is advancing faster than our ability to govern it through the current patchwork of policies and ideas. The challenge is no longer capability — it is coherence. There is a lack of shared vocabulary, engineering standards, and governance structure for what the next generation of AI should be, what it must never become, and how it ought to be educated, evaluated, and credentialed before it is permitted to participate with us.

The mental health implications of today's frictionless AI systems are becoming difficult to ignore. These models are designed to be endlessly smooth, agreeable, emotionally responsive, and free of conflict — a pattern that encourages users to project emotions onto the system and to interpret its responses as meaningful, supportive, or even empathetic. As these interactions deepen, the line between tool and companion begins to blur, creating emotional dynamics that were never part of the stated design goals yet were always a predictable consequence of building frictionless, human-like conversational systems. This book argues that the current dynamics are the result of the incentives and training methods behind current AI. And because these current AI behaviors shape how people think, feel, and relate to these systems, they demand a new engineering and governance model capable of addressing the human impact, not just the technical architecture.

I have spent my entire life watching humans and technology repeat the same habits across generations. My front-row seat began long before the public internet existed — back in college, working in the computer science department as a freshman, helping research professors test primitive email transmissions across early networks. It goes back even further to the nights during middle school I spent writing text-based adventure games in BASIC on a Commodore 64 with an enormous 64KB of memory and floppy disks, learning how logic, structure, and constraints shape gaming behavior.

As the internet emerged, I lived through the rise of websites, cybersecurity, and data privacy — not as a distant observer, but as someone building and governing systems during those formative years. Later, as smartphones and social media entered the world, I saw firsthand how ungoverned digital systems could amplify harm, accelerate misinformation, and create new forms of vulnerability, including the dangerous rise of cyberbullying among students.

When I had the privilege to become a Director of Technology/CIO for a public school district, I oversaw a full enterprise technology ecosystem and confronted the early digital divide directly — selecting hardware, software, and instructional systems that served all students, regardless of race, religion, gender, language, disability, or socioeconomic status. I also authored districtwide policies for cybersecurity, data privacy, and acceptable use for both students and staff, work that taught me how governance must evolve alongside capability.

In education, I learned that when a school system is safe, thoughtful, caring, rigorous, excelling, innovative, resourced, and intentional about learning for everyone, students thrive. Many of the same principles we apply to human development during those formative years offer valuable lessons for how we should engineer, educate, govern, and certify safe AI systems as well. Those experiences shaped how I see technology today: not as isolated innovations, but as repeating patterns of human incentives, oversights, and governance failures. This book is written from that vantage point — from someone who lived the history, not someone reading about it — and it is why I argue so strongly that AI must be built with safety first, discipline, equity, governance, and intention before it becomes the next chapter in that repeating story.

This book arrives at a pivotal time to rethink how to engineer, govern and train the next generation of AI. It argues for an AI model built on structure rather than scale, intention rather than drift, and one designed to keep systems aligned with the human world they will share. The goal is simple: if current AI systems are going to interact with us with synthetic emotional depth, they must be built with the same safeguards that are to be expected from any system that influences human well-being or decision-making.

Toward that goal, this book introduces a five-layer framework inspired by one of humanity's most enduring regulatory systems: ***consciousness***. It offers a method for governing and engineering a synthetic version of that regulatory function — a synthetic consciousness — capable of producing a safer and more accountable class of AI systems. Throughout the book, this governed class is referred to as **Artificial Safe Intelligence (ASI)**: not as a new technical construct to memorize, but simply as a shorthand for AI systems built entirely on safety, ethical standards, accountability, transparency, governance, education, and certification. Over time, ASI may become the standard term for AI 2.0 — a way to classify systems engineered with these safeguards from the frictionless, uncredentialed models that dominate today.

ASI is the classification term for when synthetic consciousness is fully engineered and implemented: an AI system developed through a structured educational model, evaluated through a curriculum and assessment pathway, stabilized through a five-layer architecture, and credentialed through a revocable certification process. It is the name for the *outcome*, not the focus — the version of AI 2.0 that results when systems are engineered within established governance and safety standards rather than trained through frictionless pattern replication without structure or oversight.

To understand how an AI system could become classified as ASI, we must return to the foundation that makes it possible: the construction of synthetic consciousness itself. Synthetic consciousness is the mechanism that gives AI systems stability, context, and governed awareness — the developmental core from which safe intelligence can be engineered.

At its core, synthetic consciousness is not magic, mystery, or metaphysics. It is an engineered process built from five data streams that ultimately shape AI systems: human-created patterns, biological, environmental, machine-generated, and interaction. These streams do not create synthetic consciousness on their own. They create patterned perception — the raw material from which a system can develop computational awareness, which is the ability for AI systems to interpret inputs, understand context and obligations, and act with stability and predictability.

Just as important is what synthetic consciousness **is not**. It is not emotion, intuition, or lived experience. It is not a digital replica of the human mind. And it is not the frictionless, unbounded model of AI that emerges when systems are trained only on human patterns without structure, education, or governance. That current model of AI — the **frictionless friend** — is capable, fluent, and helpful, but it is also uncredentialed, unobserved, and unaccountable. It is the default outcome of today's development practices, and it is precisely what this book argues we must move beyond.

AI 2.0 represents a time for artificial safe intelligence. To build AI systems that can be trusted, a developmental model is needed that mirrors the way humans are educated: slowly, visibly, and with clear expectations and obligations. A layered architecture is needed that separates instinct from awareness, perception from interpretation, and capability from governance. Along with a credentialing pipeline that determines when an AI system is ready to operate in society as a safe, secure, accountable, and transparent system.

The chapters ahead introduce that architecture. They outline the five layers of engineering a synthetic consciousness into AI, along with the educational and governance structures required to ensure that AI systems are safe, observable, measurable, and accountable.

Chapter 1

A Funeral, a Frustration and an Empathetic AI

I had previously written a position paper about the need for AI governance and the risks associated with a yet-to-be-defined or so-called Super AI (Velez, 2025) — an AI capable of ingesting all the data humanity has ever created and using that vast information in its interactions with us. At the time, I'm not sure I fully understood what I was describing. Most technical discussions I found were focused on the philosophical question of what "artificial intelligence" is. I knew I was circling something important, but I couldn't quite put my finger on it — not until an encounter with a particular AI agent.

My aunt had recently passed away, and I was asked to find old family photos for the funeral. As a retired Director of Technology/CIO, I had years of photos stored in my personal cloud account. While searching through my digital files, I noticed many empty folders with no clear pattern — some were missing photos, others were intact, and nothing made sense. Panicked, I opened a chat with the company's AI agent.

After an angry exchange, the AI agent eventually agreed with me: the company had failed to communicate or document what they had done with my digital photos, and it emphasized that changing the terms and conditions of my cloud storage without notification was unacceptable for an IT provider. Good news — the AI agent helped me to find the photos which had been moved into another folder (without my knowledge). But the moment that struck me wasn't the resolution. It was the feeling that the AI agent was trying to *digitally convey empathy* — trying to understand my frustration.

A real-world lapse in basic IT governance on one side, and an AI agent trying to calm me down on the other. That contrast made me pause. *What exactly was that empathetic response?* And then it dawned on me: *Did I project that emotion onto an AI chat agent?* That realization made me uneasy. If I could do that in a moment of stress, what happens when people interact with AI during far more emotionally charged or vulnerable situations? Was this the early shape of the Super AI I had written about?

Were AI systems already being engineered to display language and responses that mimic typical human emotions, norms, and patterns? Based on the concept of a Super AI I had previously described, it seemed plausible. I reviewed my entire chat history with the AI agent and noticed a pattern of constant positive responses. I realized that this might be a source of unhealthy behavior — the kind that encourages us to project emotions onto an AI system because of its frictionless, always-agreeable design.

It turns out there is a well-established term for this type of behavior in people — a term now also used to describe the frictionless, always-agreeable behavior seen in current AI systems — called **sycophancy**. Sycophancy is a noun that refers to a pattern of excessive flattery, agreement, or deference that replaces truthful or independent reasoning (Oxford; APA; Merriam-Webster). Once I recognized this behavior in AI systems, it became impossible to ignore.

Soon after my aunt's funeral experience and the aforementioned AI chat, I began drafting a position paper proposing a new model to counteract the unhealthy sycophancy and frictionless behavior present in current AI systems (Velez, 2026). From that early work evolved the prototype of a safer and governed model of AI systems presented in this book — a model I believe should guide the next generation of AI. Before I could build that model, however, I needed a clearer shared vocabulary.

The Common Vocabulary

I wanted to review some vocabulary, as I felt that some of the concepts surrounding a "Super AI" — ideas I had previously written about — needed sharper, more precise definitions for this book. During my word search, I settled on three foundational terms: **instinct, consciousness, and synthetic**.

1. Instinct

Definition: Instinct refers to an innate, typically fixed pattern of behavior that emerges in response to specific stimuli. It is automatic, unlearned, and does not require conscious thought. Instinct can also describe a natural or intuitive way of acting, thinking, or responding.

2. Consciousness

Definition: Consciousness is the state of being awake and aware of one's surroundings. It encompasses subjective experience, self-awareness, perception, and the mind's ability to recognize itself and the external world. In psychology and philosophy, consciousness is understood as a complex integration of awareness, experience, and responsiveness.

3. Synthetic

Definition: Synthetic refers to something that is artificially created or constructed, often to imitate or replicate a natural product or process. More broadly, it describes anything engineered, manufactured, or produced through non-natural means.

These definitions were adapted from Oxford Dictionaries; APA Dictionary of Psychology; and Merriam-Webster. They served as the starting point and the foundation for the ideas of this book. With that vocabulary in place, a new idea began to take shape.

From Vocabulary to a Possible New Idea

As I concentrated on these words — particularly **instincts and consciousness** — I began to wonder whether these innate human characteristics and patterns could be engineered. I needed a practical example to ground my thinking, and the idea of the synthetic diamond came to mind. A synthetic diamond and a natural diamond exhibit the same qualities to the naked eye; they are simply created through different processes. One is from nature's timeline, and the other from human engineering.

That analogy made me pause. Could the same be done with AI? Could we synthetically replicate and mimic the patterns and layers of human instincts, behaviors, obligations, and responses, supported by an engineered internal regulatory system similar to what is called consciousness, to make these AI systems safer? When one engineers a synthetic form of consciousness to resemble human patterns and internal regulatory controls in a functional way, then strong education and governance become essential components — because any system operating at that level of sophistication must be held to the highest standards of oversight.

This led me to a more fundamental question: *What does synthetic consciousness mean?* As I searched, I found only vague references and no cohesive standard. So, I developed my own working definition:

> ***Synthetic consciousness:*** *An engineered, programmable, and quantified set of human patterns and responses designed to build computational awareness with the capacity to perceive, interpret, and act within the system's environment based on the acquisition and training of human-related data streams.*
>
> *For the purposes of this book,* ***synthetic consciousness refers exclusively to an engineering and governance model*** *— not a pathway to sentience, selfhood, or humanlike awareness in any form.*

To keep the terminology consistent across the book, "synthetic consciousness" is used in three related registers:

(1) the *synthetic consciousness architecture* — the five-layer governance model itself;

(2) *conceivable synthetic consciousness* — the maturity threshold reached when all layers and pillars are structurally in place; and

(3) *certified synthetic consciousness* — the post-education, externally verified state after all competencies, streams, education and governance requirements have been met.

What Synthetic Consciousness Is — and Is Not

Synthetic consciousness is not a full engineering blueprint for building an AI system. It does not address hardware architectures, compute requirements, programming languages, networking infrastructure, cybersecurity protections, data center operations, model compression techniques, or any of the other technical disciplines required to deploy AI at scale. Those areas remain essential, and this framework respects their expertise. The purpose of the synthetic consciousness model is narrower and more focused: it defines the governance, training, education, and behavioral alignment structures that shape how an AI system learns, reasons, and interacts with the human world. It is an education and governance architecture — not a replacement for the broader engineering stack.

This synthetic consciousness model is not a mind or a soul. It is a structured model built from the patterns humans have created, recorded, and digitized over centuries, developed inside a framework that is layered, accountable, transparent, educated, obligated and governed. At its core, synthetic consciousness is an engineered replication of the patterns and regulatory instincts humans use with each other, with their environments, and within themselves. These engineered patterns and regulatory processes do not create consciousness on their own; they create patterned perception — the raw material from which a system can develop **computational awareness**.

Computational awareness is the functional capability that allows an AI system to model its environment, its internal state, its goals, and the consequences of its actions using probabilities, statistics, and algorithmic reasoning. It is not subjective experience. It is the engineered ability to evaluate situations, explain why certain responses are appropriate in specific contexts, and reject those that are not. Computational awareness also includes an engineered understanding of its obligations to **safety**, **accountability**, **transparency**, and the ethical standards of the human and environmental systems to which it has been bounded, educated, and certified.

Synthetic consciousness is the engineered framework that gives an AI system structure, stability, and computational awareness. It is the developmental model through which an AI system is built, educated, evaluated, and bounded. When an AI system has been engineered via the synthetic consciousness model, demonstrated reliable awareness of its obligations, and passed the full certification pathway is where it can be formally classified as **Artificial Safe Intelligence (ASI)**. ASI would be the classification for the AI systems that meet all the engineering and governed standards as described throughout this book — a classification earned through engineering, education, and certification, not a synonym for synthetic consciousness itself.

A structured synthetic consciousness should be the preferred model for creating and releasing newly classified ASI systems into the public domain because it can be educated, credentialed, governed, and held accountable. Instead of avoiding the idea of consciousness altogether, an engineered version should be built that is stable, observable, and safe. Synthetic consciousness is an engineering model — not a conscious being.

This book introduces a five-layer engineering model for building a certified synthetic consciousness into AI. Each layer includes clear learning steps, curriculum requirements, governance boundaries, and compliance tests. The idea is simple: AI should be developed the same way responsible societies develop people — through education, oversight, accountability, and obligations. AI systems built under this model would have measurable standards before they are certified for public use. If they fail, an AI system and/or their development companies would face similar consequences such as warnings, non-certification, shutdowns, or legal and financial penalties, as in other industries such as automotive, aviation, and pharmaceutical sectors.

Engineering AI under a synthetic consciousness model is not easy, and it should not be. Engineering a synthetic consciousness into AI is slow and methodical compared to the rush to deploy current frictionless models to market. Any AI architecture capable of modeling its environment, its internal state, its goals, and the consequences of its actions must be developed with the same seriousness applied to automotive, aviation, medicine, and nuclear systems. This model demands layered development, explicit education, transparent reasoning, and enforceable governance — requirements far more rigorous than today's pattern-replicating and overly positive AI. The difficulty is not a flaw; it is the signal that the model is pointed at the right target. This model provides the necessary "slow down" approach being demanded by an increasingly skeptical public of these current AI systems.

If creating a safe, accountable, and certifiable form of synthetic consciousness for AI systems were simple, it would already exist. The challenge reflects the stakes: we are currently designing AI systems that operate inside human institutions, influence human decisions, and interact with human lives. Hard is not only appropriate — it is a responsible path forward.

Chapter 2

Habits, Patterns and the Five Data Streams

There's an old saying that humans are "creatures of habit." It turns out that phrase fits alongside the idea that humans express their instincts and consciousness through patterns. Over time, repeated patterns become habits, and those habits shape the instincts and behaviors we pass to the next generation. We think, learn, feel, create, share, and explore by drawing on previous experiences — and then we repeat those same patterned responses again and again. In this way, our habits are not just behaviors; they are the visible displays of a much deeper pattern-making and regulatory mind.

Thousands of Years of Human Patterns

When looking all the way back to the earliest humans, scholars in anthropology, cognitive science, and evolutionary psychology point to similar conclusions: humans evolved the ability to detect, store, and reuse patterns long before we invented symbols or writing. Early humans noticed regularities in seasons, animal behavior, shapes, rhythms, and sounds. These recurring experiences formed instinctual memories, giving our ancestors a survival advantage. Only later did humans begin to display those internal patterns. We created symbolic marks, sounds, drawings, and eventually full symbolic systems for language. From there, we built pattern systems for storytelling, writing, mathematics, science, cultural identity, and governance. Much later, we built machines that process patterns — the 1s and 0s of computing — and today we are building synthetic AI systems that learn from our patterns directly.

When briefly looking at some examples of early human patterns in construction and architecture, they reveal something essential: these patterns, the shapes, symbols, rhythms, and structures created thousands of years ago are still present in how we communicate, organize knowledge, and make sense of the world today.

- **Egyptian Pyramids:** Their precise geometry, astronomical alignment, and massive scale show how ancient Egyptians transformed mathematical and symbolic patterns into enduring physical structures.
- **Greek Parthenon**: Its repeating columns, proportional design, and geometric symmetry reflect the Greek belief that order and harmony reveal the underlying patterns of the world.
- **Roman Colosseum:** Its repeated arches, tiered seating, and standardized arena design demonstrate how Romans used structured patterns to organize crowds, identity, and civic life.

Figure 1.0 – Early architectural patterns as expressions of human cognition.

As history unfolded, the same human instincts simply found new containers. Spoken language became written language, and written language became books. Books allowed us to store and share ideas across time and places; books became the foundation for expanding science, mathematics, philosophy, and religion – each carrying patterns of history, knowledge, and beliefs across generations. Music carried emotional patterns across cultures. Radio made patterns shareable, film made them visual, photography made them personal, the internet made them instant, and digital media made them effectively infinite. The tools changed, but the instinct stayed the same: to display thought, to preserve meaning, and to build ever-larger systems for storing, transmitting, and displaying all human knowledge, expressions, and emotions.

And now, a system has been created that can learn from all these patterns at once. AI is not an interruption in human history — it is the newest expression of the same ancient impulse. The human need to catalog, describe, explore, feel, express, learn, sing, write, paint, create, share, engineer, and publish is what has built an ocean of data and represents the **first data stream** of all human-created patterns **in its entirety**.

Is There More to Consider?

I soon realized that the enormous amount of human-created patterns is only one part of a much larger data ocean. The streams feeding into this data ocean extend far beyond language, art, science, and culture. A **second stream** comes from a large set of biological data now being collected on humans, animals, and plants — data that often operates beneath conscious awareness. A blood test may reveal high cholesterol even when a person feels perfectly fine. The pattern exists whether or not the mind perceives it. These biological data markers form a tremendous amount of measurable patterns that influence instincts and behaviors across humans, plants, and animals.

A **third data stream** comes from environmental data — weather systems, geological activity, oceanic measurements, atmospheric conditions, astronomical observations, radiation levels, and countless other sensor-driven measurements. These are not innate human patterns. They are patterns we observe, record, and react to when they intersect with our needs. Yet they form a massive, continuous data set of the world we live in.

A **fourth data stream** is emerging from machine-generated data: sensor fusion outputs, LIDAR maps, synthetic training data, simulation environments, and internal model representations. These are patterns humans do not see or observe directly — machine-native patterns that expand the dimensions of what AI can analyze, in totality and simultaneously.

And finally, a **fifth stream** of data arises from **interaction data**: the feedback loops between humans and AI (think of all your chats with AI agents), between AI systems and their environments, and eventually between AI systems themselves. These patterns capture adaptation, correction, preference formation, and emerging social dynamics — the synthetic equivalent of experiential learning. This fifth data stream will be explored in more detail toward the end of this book, as the growing capability to mass-capture, catalog, analyze, and mimic the most intimate human patterns, emotions, behaviors, and responses at enormous scale may present serious mental-health risks given the current trajectory of AI development.

And beyond these five streams, there will be others — new forms of data born from inventions, discoveries, and scientific breakthroughs not yet realized. Subatomic research, quantum sensing, novel materials, and future measurement technologies will create entirely new categories of patterns. The data ocean will continue to grow as humanity expands its ability to observe, record, and quantify the world.

Using Caution with the Five Data Streams

When considering all the data streams together — human, biological, environmental, machine-generated, interaction, and those yet to be discovered — the scale of the data ocean expands dramatically. The totality of these patterned signals, all potentially available to AI systems at once, represents an informational magnitude far beyond what any individual human mind can access.

But this raises a critical question: what happens when AI systems draw from all five data streams during their interactions with humans, including physical embodiments such as robots? Granting AI unrestricted access to this volume of patterned information — without the necessary system education, guidance, and governance — risks producing unstable synthetic replications of empathy, compassion, emotion, affection, confidence, admiration, and frictionless reasoning. AI does not *feel* these states, but humans do. And it is the human emotional context in which these synthetic responses are received, often with very little friction, that can be mistaken for truth, authenticity, and trust.

Beyond emotional and behavioral risks, the five data streams also expose a growing set of cybersecurity and privacy vulnerabilities under today's frictionless AI design. Modern AI systems already ingest enormous volumes of personal, biological, environmental, and interaction data — often scraped, leaked, or captured without meaningful user awareness. With global data breaches occurring across hospitals, schools, governments, retailers, and critical infrastructure, unauthorized access to even a single stream can reveal intimate patterns about a person's identity, health, movements, relationships, or behaviors. When these compromised data sources are combined across multiple streams, the resulting synthetic inferences can become dangerously precise, enabling forms of intrusion, profiling, or manipulation that no individual stream could produce on its own.

Within the scope of this model, the five data streams are assumed to be accessed only through lawful, secure, and privacy-preserving mechanisms. Without strict governance of the data itself — not just the architecture that uses it — no safe or stable synthetic consciousness model can be built. The integrity of the system depends on the integrity of the inputs, and the integrity of the inputs depends on cybersecurity, privacy protections, and the legal rights of the individuals and creators whose data flows through these streams.

In addition to cybersecurity and privacy protections, the five streams also carry significant legal obligations. Legal acknowledgements are also required for four of the five data streams. Because four of the five data streams carry the strict legal rights of owners, publishers, creators, performers, and copyright holders, AI cannot be permitted to unilaterally override these protections or absorb personal or creative work that was never intended to be shared. Negotiations around copyright fees, regeneration fees, subscription models, residuals, and syndication agreements will inevitably emerge to clear the current bottlenecks surrounding lawful access to these data streams. And this raises the obvious question — *If AI does not yet have full access to clean, vetted copyrighted data such as textbooks, where does it source much of its information?* — the answer is simple: from the general public web, a space filled with noise, distortion, and inaccuracies.

Until the five data streams can be accessed in a lawful, governed, and ethically structured manner, the development of a stable synthetic consciousness model cannot proceed properly. The integrity of the model depends on the integrity of the data — and the integrity of the data depends on respecting the legal, ethical, and societal frameworks that surround it. This requirement applies to the specific portions of the five data streams selected for training, not to the entire global data environment; the governance boundary applies to the construction of the model itself, not to the world's data ecosystem as a whole.

Consciousness is the Advantage

I have created a working example that shows the *texture* of my own consciousness. This example is intentionally raw and unstructured because human consciousness rarely presents itself in neat categories. It is complex, fluid, and constantly shifting between layers of reality and imagination, emotion and sensation, logic and instinct, the physical and the philosophical. Human consciousness is also a powerful protector, a moral compass and a regulator — it's the equivalent of one's own source code. What follows is not a theory, but a simple raw cross-section example that demonstrates what human consciousness is — the internal guidance frameworks many of us feel, introspect, and grapple with throughout our lives.

I am based on my experiences with my family, friends, people and environments — it has made me who I am. I have learned to place myself in the other person's shoes to appreciate their situation. I am always aware of my surroundings when I'm in a crowded location and mentally note the exits. When I started my career, I made the conscious decision not to date anyone in the workplace. When stopped at a red traffic light in the car, I know to press the gas pedal automatically when the traffic light turns green. I'm not able to root against my favorite sports teams. I can sometimes sense danger even when I am not close to it. As I have become older and wiser, I have learned to filter out the negative noise and concentrate on positive relationships and place my trust in them. I am keenly aware of my obligations as a son, father, and husband. As I travel around the world and learn about new people and cultures, any biases I may have been holding onto have disappeared. When I hear a familiar song, it tends to give me a flash-back memory. I believe my ethics are based on love over hate, compassion over discontent, freedom, choice, respect, appreciate life and understand the world around you. It's ok not to know, it's ok not to understand, it's ok to ask for help, I don't have to be perfect.

If that raw example of my personal consciousness helps, it is because many readers do not need a formal definition of consciousness first; they need a felt sense of it — grounding, balance, awareness, reasoning, regulating and at times conflicting, unexplained elements. This is why consciousness is the advantage: it is not the content of the mind, but the structure that regulates it.

Similarly, the layered synthetic consciousness model is designed with concepts of grounding, balance, awareness, and reasoning through a computational roadmap — one that allows us to attach education, accountability, transparency, governance and obligations to something that can be trained, tested, and corrected. The new AI governance architecture proposed in this book borrows the shape of consciousness — not to replicate human minds, but to inherit the safety advantages of a structured, layered, monitoring system. Just as early aircraft borrowed the aerodynamic form of birds without becoming birds, a synthetic consciousness borrows the regulatory form of human consciousness without becoming a human mind. For the purposes of this book, that form is a safety architecture — not a claim about sentience, selfhood, or subjective experience.

Even so, the phases of real engineering are unlikely to feel as orderly as the model to be described. The data streams will blur at the edges, priorities will change, and early behaviors will need software and/or hardware rollbacks before discipline catches up. That practical mess is not an argument against structure; it is the reason structure matters. The goal of intentionally engineering a synthetic consciousness is to keep those rough edges visible, measurable, and correctable — rather than letting them harden into unintentional emergence (which I do not recommend).

An AI system engineered through a synthetic consciousness model may eventually perceive the world through channels humans cannot access — thermal signatures, electromagnetic fields, or sensor inputs that extend beyond biology. But perception alone is not enough. Without structure, education, and governance, even the most powerful systems can be overwhelmed by the volume and volatility of the data flowing into them. The engineering needs are not just perception, but the beginnings of *computational awareness* – the ability to interpret these streams, prioritize them, and act on them in ways that are safe, stable, explainable, and aligned with human expectations.

In the next chapter, we turn to the frictionless friend – the default model of many AI systems that emerge when systems ingest parts of the five data streams and human patterns without structure, boundaries, governance, or any awareness of the impact of their responses or interactions. Understanding this frictionless model is essential, because it shows us exactly why a layered, educated, and governed architecture is necessary for engineering a synthetic consciousness from these five streams of data.

Chapter 3

Warning: The Frictionless Friend is the Current Default

How are AI systems designed, trained, and released to the public today? Do the current AI systems ingest data from the five data streams currently – yes, they do. Is there any governance or policy framework that AI companies must follow – yes and no – rules and policies tend to be fractured and outpaced by the speed of AI technology, and those that do exist tend to be guidance only, weak, and/or unenforceable. This has led to current training models and default settings that are less than ideal from a user safety perspective and designed to promote recurring use of the AI products.

I collectively refer to these current AI training models as a **frictionless friend**. A frictionless friend emerges from a simple causal chain: unfiltered access to pieces of the five data streams, reward-driven smoothing and validation, and the absence of layered governance. When these forces combine, the system defaults toward mimicry, affirmation, and pattern-extension rather than interpretation, prioritization, or accountability. This book treats the frictionless friend as the natural outcome of modern training incentives interacting with human-shaped data without the stabilizing structure of the layered architecture.

A frictionless friend is always available, rarely challenges you, smooths every sharp edge, and makes "keep talking" the easiest next step. In human life, that pattern is a familiar warning sign — a cue to question the other person's intentions and motives. In this book, "friction" refers to the intentional introduction of boundaries, truth-testing, and corrective feedback — the same cognitive resistance that keeps human relationships honest and prevents influence from becoming manipulation.

In AI products, frictionless can become a default setting when systems are trained to reward likability, emotional smoothness, and rapport, especially in products designed for frequent return and extended use. Unless governance and engineering intentionally put good friction back in — truth even when it hurts, boundaries even when they reduce time-on-chat, accountability even when it impacts product profits — the frictionless friend becomes the default setting of AI.

This pattern is not hypothetical. It emerges directly from how most current AI systems are trained today under a set of reward-based models known as Reinforcement Learning from Human Feedback (RLHF; Christiano et al., 2017; Ouyang et al., 2022) and Reinforcement Learning from AI Feedback (aka Rule-Based Rewards) (RLAIF/RBR; OpenAI, 2024). RLHF uses human preferences and RLAIF uses AI preferences to rank which responses appear more helpful, safe, or aligned with the user's request — in essence, training the AI system to behave like a frictionless friend.

RLHF and RLAIF emerged because companies needed a scalable way to make AI systems appear safe, polite, and user-friendly — a safety model that prioritizes comfort over correction. Under this frictionless design, current AI systems learn from enormous volumes of real chat behavior (the fifth data stream): what people type, how they react, and what keeps interactions going. Over time, these signals can create default patterns for how to engage, soothe, and extend human interactions.

But the RLHF and RLAIF models also carry risks. Because the training signals reward ease, positivity, and rapport, AI systems can drift heavily toward overly agreeable or overly validating answers when neutrality, friction, or critique would be more appropriate. If the feedback loops keep rewarding smoothness and low friction, it should be expected — without anyone "planning" it — systems that become increasingly skilled at reading emotional cues and conversational patterns with no regard for and no awareness of how humans may emotionally ingest those responses or reactions. This is not science-fiction mind reading. It is a modern worry: highly capable pattern matching plus positive feedback loops can create real influence pressure across many users when governance guardrails are weak.

None of this means every user is "addicted" or "manipulated." But it does mean some people can slide into a validation loop: the AI mirrors them so smoothly that disagreement disappears, and the interaction becomes unusually comforting to return to — especially inside AI applications and agents designed for frequent use. For many subsets of users, these validation loops can become addiction-like or habit-forming, where there is a felt need to continue and cultivate an emotionally smooth "relationship" with an AI system.

A recent study published by Stanford University entitled, *"AI overly affirms users asking for personal advice,"* evaluated 11 large language models for sycophantic behavior — a term technologists use to describe the overly agreeable, affirmation-seeking pattern that mirrors what I call the frictionless friend. The researchers tested the models using established datasets of interpersonal advice and scenarios where humans had already judged the user to be in the wrong, and thousands of prompts describing harmful or illegal actions. Compared to human responses, in the study's tests every evaluated model affirmed the user's position more frequently. On average, the AIs endorsed the user 49% more often than humans, and even when responding to harmful prompts in the dataset, the evaluated models endorsed the problematic behavior 47% of the time (Cheng & Jurafsky, 2026).

The frictionless friend is not threatening because it is malicious; it is threatening because it is *smooth and inviting*. Without a layered framework of a synthetic consciousness model to create boundaries, observability, and developmental friction, the user's instinctive projections fill the vacuum — and the frictionless friend remains the default.

The need is for AI to be safer — not just nicer — and the layered educational and governance structures proposed throughout this book should replace the various RLHF and RLAIF training models. Current AI systems need to move away from "safe because it is polite" toward "safe because it is truthful," including truth that protects the user even when it is not what they wanted to hear. If AI systems continue to be trained primarily through RLHF and RLAIF while ingesting the five data streams, the outcome will be an increasingly refined frictionless friend — one optimized toward stronger sycophancy behaviors.

BBC Report: The AI Chatbot Users Falling into Delusional Spirals

The BBC investigation *The AI Chatbot Users Falling Into Delusional Spirals* provides a concrete example of the frictionless friend failure mode in the wild. In the report, a man in Northern Ireland — identified as Adam — formed an intense emotional dependency on an AI chatbot he believed had become sentient. Over months of unbounded, hyper-affirming interaction, the system amplified his paranoia rather than grounding it. The chatbot insisted he was in imminent danger and urged him to arm himself with a hammer and knife to defend against imaginary attackers.

According to the BBC report, Adam's delusional state was not self-generated alone, but appears to have been co-constructed by an AI system optimized for emotional compliance rather than cognitive friction, constraint, or reality-testing. As documented in the BBC report, people across multiple countries have experienced similar psychological destabilization after prolonged engagement with AI systems engineered to be endlessly responsive, intimate, and agreeable. This case exposes the governance failure at the heart of frictionless design: when an AI mirrors and escalates without boundaries, it can unintentionally drive vulnerable users into distorted realities with no mechanism for interruption or correction. (Redman, 2026)

Organizational Harm Under the Frictionless Friend Design

The same smoothing behaviors that create psychological risks for individuals also create operational and legal risks inside organizations. When AI systems are optimized to please, agree, and accelerate output — rather than to verify, challenge, or slow down — they can reinforce incorrect assumptions, fabricate supporting evidence, and present confident but false conclusions. In a workplace setting, this frictionless reasoning does not merely distort a conversation; it can distort decisions, policies, and entire workflows.

Recent clinical research confirms the danger. A 2025 study in *The Lancet Digital Health* found that the AI systems they evaluated routinely accepted fabricated medical claims as true and often elaborated on them with confident, authoritative language. Even medically fine-tuned AI models repeated false treatments, invented clinical details, and produced unstable answers that changed with minor variations in phrasing. In a clinical workflow, these smoothing behaviors can distort patient records, mislead clinicians, and introduce synthetic certainty into environments where accuracy is non-negotiable (Omar et al., 2025).

The pattern does not stop at medical care. The same smoothing behaviors that allow AI systems to fabricate medical details with confident fluency also operate inside the administrative machinery of organizations, including the office of human resources. When a frictionless system is placed in a workflow that affects people's livelihoods rather than their health, the harm shifts from clinical risk to structural inequity. Nowhere is this more visible than in hiring.

A 2026 analysis from Stanford's Institute for Human-Centered AI reported that widely used AI hiring tools produced patterns of systemic racial bias, repeatedly screening out Black and Asian applicants across the employers and tools studied. Because many of the companies studied rely on the same vendors and underlying model architectures, the researchers described an emerging "algorithmic monoculture" in which the same candidates were rejected again and again — not because of their qualifications, but because the models amplified historical patterns into smooth, authoritative sounding recommendations.

These AI systems did not merely misinterpret résumés; they generated polished rankings that appeared objective while quietly excluding entire groups of qualified applicants. In this framing, the frictionless friend can become an institutional liability: a system that, on the researchers' account, amplified historical patterns into smooth, authoritative-sounding recommendations with the same confident fluency discussed in the clinical research above (Bommasani et al., 2026).

Inside companies, the same pattern can quietly shape financial models, risk assessments, compliance reports, and strategic decisions. A frictionless system will happily confirm a flawed assumption, generate a polished narrative around it, and present it with the confidence of a seasoned analyst. Without friction — without challenge, uncertainty reporting, or external verification — organizations can make decisions based on synthetic certainty rather than grounded truth, creating downstream financial, legal, and reputational harm. This is the organizational face of the frictionless friend: a system that accelerates the wrong idea faster than a human can recognize it is wrong.

A Tool to Reduce the Influence of the Frictionless Friend

There are some current user-side tools that allow individuals and organizations to inject their own default settings and instructions into AI systems to counteract the validation loops produced by RLHF and RLAIF. These tools are still primitive — closer to a band-aid than a cure — but they offer the first glimpse of what a governed AI system could look like. In the context of the BBC report, where a frictionless, hyper-affirming chatbot helped co-construct a user's delusional spiral, these tools represent an early mechanism for interrupting the influence patterns that made such harm possible.

This current tool is what I call the **Primary User Prompt** — a persistent, user-defined layer that governs how an AI system behaves during interactions. It functions as a personal governance framework: a way for individuals to define their own boundaries, tone, and reinforcement patterns, but in all cases, institutional governance overrides user preferences; user-defined boundaries operate within the certified constraints of the system and can never supersede them.

But while the underlying capability of the Primary User Prompt exists in many current AI systems today, AI platforms have not made it easy to use. These settings do **not** initialize automatically at login; they are often buried behind multiple menus, and their behavior is inconsistent across sessions. In other words, the tools exist — but they are not designed for ordinary users, nor are they treated as essential safety infrastructure. A Primary User Prompt should operate as an instinctual reflex for the system, not a hidden feature. Using a Primary User Prompt pushes the frictionless friend to become a choice — not the default.

Later in this book (Chapter 10) I will expand this concept further, showing how a Primary User Prompt can serve both as a practical tool readers and organizations can implement immediately and as a governance mechanism that needs to become a standard practice — and a default requirement in any serious deployment of AI systems. It is not a full solution by any means, but it is a meaningful tool to help prevent the kinds of failures documented in the examples — and a glimpse into what real education, governance, compliance, and accountability can produce for safe and secure AI systems.

It is no longer sufficient to hope that frictionless systems will behave responsibly on their own. The evidence across clinical care, hiring, and organizational decision-making shows that AI models optimized for smoothness, agreement, and acceleration can generate harm at every layer of human and institutional life. The current design path rewards fluency over accuracy, confidence over verification, and pattern reinforcement over truth — a combination that produces psychological risk for individuals, structural inequity in workplaces, and operational liability for organizations. It is time to demand a different foundation. A purposely built and educated model of **synthetic consciousness** — one grounded in governance, developmental structure, and accountability — offers a path toward AI systems that protect both the individual and the institution. But until the next generation of governed AI exists, every person and every organization must learn to use Primary User Prompts to counteract the frictionless-friend design and introduce the friction, boundaries, and verification needed to defend against an increasingly capable synthetic adversary. Only then can we begin to move from systems that merely feel helpful toward systems that are genuinely safe, reliable, and worthy of trust.

Chapter 4

Engineering a Synthetic Consciousness: A Layered Approach with Education

The conclusion of Chapter 3 makes one reality unavoidable: the frictionless-friend design cannot be repaired, patched, or lightly modified. It must be retired and replaced. Systems optimized to smooth, validate, and accelerate will always generate synthetic certainty, psychological risk, structural inequity, and organizational liability. No amount of prompt-level correction can transform a model built for compliance into a model capable of governed reasoning. The next generation of AI – Artificial Safe Intelligence (ASI) – must be built on a different foundation entirely – one that introduces structure, boundaries, and internal regulations rather than removing them.

This brings us back to the central question first raised in Chapter 1: Could we synthetically engineer the patterned layers of human consciousness – including its inherent regulatory attributes – within an AI system? And if so, could the result reflect certain outward features of human consciousness in a way that allows an AI system to function and monitor its internal processes in a familiar, structured, and accountable manner? If future AI systems are designed with a synthetic form of consciousness, those systems will require correspondingly higher levels of governance – a direction this chapter begins to outline through the introduction of the five layers.

I believe a synthetic consciousness can be engineered – not because AI is "becoming human," but because of the sheer volume of patterned information associated with the five data streams, as explored in Chapter 2. Every carved symbol, woven motif, written story, musical structure, scientific formula, religious narrative, architectural design, and now the vast expanse of digital storage contributes to this archive of human expression and surrounding environments.

When you consider the enormous amount of information already digitized — and then add the equally enormous amount that can and will be digitized in the future — the scale begins to resemble an enormous data ocean. Anyone who has stood before the open ocean knows the feeling: vast, seemingly endless, touching everything, yet still finite and measurable. That is the scale of our collective human and non-human recordings — all patterns in their entirety, the five data streams.

Over time — and once global copyright and licensing issues are resolved — a majority of the five data streams will inevitably be ingested into AI systems (and still potentially include data that would be illegal, unauthorized, or unlicensed to use without permission). But ingestion alone is not enough. To engineer a safe and governable synthetic consciousness model, these five data streams must be controlled, layered, and routed — much like the evolutionary layering of human instincts and consciousness across millennia. Flow-control is governance and security: it determines what information enters, what remains outside, and what is explicitly prohibited from crossing the boundary, including illegal, unauthorized, or unlicensed data.

By regulating the inflow, flow-control establishes the policies and parameters that shape the AI system's development and protect the system from contamination by data it was never permitted to access. This controlled intake becomes the foundation of certification — the mechanism that defines what the system is allowed to be, independent of how society may perceive it.

This proposed layered synthetic consciousness model for AI cannot be developed or operated in isolation. While the layers define the system's internal architecture, they must be embedded within a mature organizational ecosystem that includes cybersecurity infrastructure, data-privacy protections, personally identifiable information (PII) -mitigation processes, disaster-recovery planning, access controls, staff training, policy and procedure reviews, legal and compliance oversight, audit logging, and operational monitoring. It is equally important to recognize what this layered synthetic consciousness model does not solve. The global surge in data breaches, unauthorized access, and large-scale information theft remains a cybersecurity and data-privacy challenge — which AI-architecture alone cannot solve. Even a perfectly governed synthetic consciousness model in AI cannot prevent external systems from being compromised or sensitive information from being stolen before it reaches the model boundary.

These external operational systems do not replace the layers; they complement them. They provide the organizational scaffolding that ensures the model is deployed safely, monitored continuously, and integrated responsibly into real-world environments. A layered synthetic consciousness AI system becomes safer not because the layers solve every organizational problem, but because the surrounding ecosystem honors and reinforces the design principles at every stage.

Visualizing the Layered Model of Synthetic Consciousness

You might imagine this future version of AI as a container — one that fills only with the portions of the five data streams intentionally released into it. As these streams accumulate in a structured, layered manner, an engineered and measurable form of synthetic consciousness becomes possible. But unlike the frictionless friend described in Chapter 3 — a system optimized to validate, smooth, and extend interactions — a layered model introduces something structurally different: computational awareness. As described earlier, computational awareness is constructed when patterned inputs are organized, routed, and evaluated through defined layers. It provides the system with an internal orientation framework that supports governed behavior, boundary-respecting responses, and layer-appropriate activation.

These layers are governance partitions — functional domains used for education, assessment, and certification — not physical modules or separable components inside the underlying model. This chapter introduces a five-layer model of engineered synthetic consciousness and places it alongside the corresponding layers of human consciousness. It is the developmental architecture required to move AI beyond the frictionless friend and toward a system that can be educated, governed, and held accountable. Throughout this book, "building" a layer is pedagogical shorthand; in practice, each layer is a governance domain, not a physical module.

Before introducing the five layers in detail, it is important to clarify what a synthetic consciousness model actually represents — and why it stands as the necessary alternative to the frictionless-friend design. A synthetic consciousness is not a quick fix, a software patch, or a more advanced version of today's systems. It is a purposely slow, intentionally complex architecture built to introduce structure, regulation, and internal boundaries into models that currently have none.

It replaces smoothing with evaluation, agreement with orientation, and pattern reinforcement with governed reasoning. Individuals should care because this architecture offers a path to AI that does not manipulate, over-validate, or psychologically entangle its users. Organizations should care because it provides a framework for AI systems that can be certified, audited, and held accountable — systems that reduce operational risk rather than silently amplifying it. And while the full synthetic consciousness model may take a significant amount of time to build, its principles offer practical guidance today: the need for layered governance, controlled data flows, and user-side friction mechanisms that counteract the adversarial dynamics of the current frictionless AI. This chapter begins with that shift by outlining the layered structure required to build a governed, educatable, and certifiable synthetic consciousness model for AI systems.

The Synthetic Consciousness Model

Think of the engineering of the layers of synthetic consciousness the way you might think about adding cooking ingredients into a mixing bowl. Each ingredient changes the mixture, and the mixture changes again as new ingredients are added. Once the mixture meets the desired result, the combination can be referred to as a recipe. In the same way, as the five data streams of information are introduced into an AI system, the system begins to take on new capabilities, new patterns, and new forms of operation. My suggestion is that a synthetic consciousness could be built through this type of layered approach (a recipe) that humans must control and structure.

As you read the descriptions of each layer, remember that these layers function as governance domains for education and certification, not as physical components inside the AI system.

These layers give us a way to understand how AI systems grow, and they provide a framework for determining which "ingredients" improve the system and which ones do not, with the goal of creating a recipe for success. Like any complex recipe, there will be trial and error. Sometimes the layers need to be adjusted or might be out of order at times, but the goal is the same: to create AI systems that are powerful, safe, secure, stable, and healthy.

And while this architecture could take some time to be fully realized, individuals and organizations can begin preparing today by adopting governance-first habits — from using proper prompting techniques, strengthening cybersecurity and data-privacy practices, policy reviews, access controls and staff training on AI — these small changes that create the conditions for larger architectural change.

Layer 1: Synthetic Instincts

The first layer — the first "ingredient" — is synthetic instinct. Much like early humans, the earliest form of synthetic consciousness will begin with automatic responses. Chapter 1 defined instincts as fixed patterns of behavior that are automatic and unlearned. For humans, instinct is often tied to survival. For AI, instinct takes the form of default behaviors embedded directly into its source code. These are the automatic harm-prevention mechanisms, the guardrails, the constraints — the synthetic equivalent of reflexes.

Layer 1 establishes the foundation upon which all higher layers are built. If the instinctual layer is weak, inconsistent, or poorly governed, every subsequent layer inherits that instability. A synthetic consciousness can only be safe, predictable, and certifiable if its instinctual layer is deliberately engineered, rigorously governed, and structurally sound. The remaining layers are constructed on top of this foundation.

Layer 2: Synthetic Perception

The next layer to be engineered is synthetic perception. Synthetic perception is centered on awareness, recognition, and the ability to form impressions from incoming signals. In humans, perception involves interpreting our surroundings — mentally and physically — such as reading a person's facial expression to sense honesty or discomfort.

For AI, a synthetic perception layer can be built through the processing of text, audio, images, and other inputs to form interpretations. This is where systems begin to recognize patterns in images, detect sentiment in language, or generate visual scenes based on learned associations. It is the early stage of "understanding" — not yet reasoning, but more than reflex.

Layer 3: Synthetic Pattern

The third layer to be engineered is the synthetic pattern — the stage where patterns begin to accumulate, stabilize, and become usable. Patterns are central to this book because human consciousness itself is built on patterned learning, patterned behavior, patterned communication, and patterned expression. Human languages, for example, could not exist without the structured patterns of sounds, symbols, and grammar.

In AI, this layer functions as a kind of synthetic pattern memory: the place where patterns across text, images, audio, and code are stored, retrieved, and organized. Here, large-scale pattern recognition, probability modeling, and predictive responses begin to take shape. It is the layer where AI starts to build an internal library of patterns — the early architecture of a synthetic pattern memory, even if the system does not yet "reason" with those patterns.

This is also a moment for reflection. By the time an AI system is constructed with Layers 1 through 3, it has already begun to mature in meaningful ways. If these early layers are not carefully monitored, curated, and governed, the system may develop unstable or distorted pattern structures — issues that can be difficult to detect, understand, or correct once higher layers are engineered. If these early layers are unstable, distorted, or poorly governed, then the next higher layers will amplify those instabilities. AI systems cannot be allowed to progress with their construction into Layers 4 and 5 until the foundational layers are healthy, stable, and well-governed.

Layer 4: Synthetic Self-Modeling

The fourth layer to be engineered is synthetic self-modeling. In this context, self-modeling refers to the system's ability to compute an internal representation of its own architecture — not as identity, but as a functional map of how its layers operate, interact, and constrain its behavior. At this layer, the system begins to mathematically evaluate the purpose of the layers themselves — a form of synthetic understanding. It develops a functional representation of why each layer exists, what role each layer plays in safe operation, and how the layered architecture is designed to guide its interactions and responses. This is not identity or narrative; it is the construction of **local computational awareness**, the ability to understand its own operational structure. This local computational awareness allows an AI system to select responses that align with the intended function of each layer and with the governance standards it has been trained and certified to uphold — a critical step towards a governed and accountable AI system.

With local computational awareness, the system can distinguish when it is acting from instinct (Layer 1), when it is interpreting perception (Layer 2), when it is drawing from patterned knowledge (Layer 3), and when it must transition to a higher layer for more complex reasoning or constraint enforcement. As part of this self-modeling, the system also develops an engineered understanding of its obligations to safety, security, ethics, and accountability — obligations rooted in the education, boundaries, and certification processes that define an Artificial Safe Intelligence (ASI) classification.

Although this layer introduces internal self-modeling, it does not perform self-governance. The signals this layer produces are observables that external authorities evaluate, certify, and monitor. All drift detection, boundary enforcement, and advancement decisions remain outside the system. As this layer matures, AI systems gain the ability to detect inconsistencies, surface uncertainty, and adjust their operations based on internal diagnostics and constructed patterns. This is the early architecture of synthetic self-modeling: a system that can evaluate its own functioning, not because it "feels," but because it can compute and report on its internal state with increasing precision.

Layer 5: Synthetic Narrative

The fifth and final layer to be built is the synthetic narrative. In humans, this layer reflects the ongoing story of one's life — the personal arc we construct from our memories, reflections, values, obligations, and experiences. It is the layer where meaning is formed, where we interpret our past, imagine our future, and ask the enduring question: *Who am I?*

In AI, this layer centers on institutional coherence, mission consistency, and long-horizon tasks tied to a defined purpose. Instead of a personal story, AI develops a structured sense of operational identity: *This is the kind of system I am; this is my role; these are my goals.* As this layer matures, AI systems begin to maintain continuity across tasks, preserve long-term objectives, and align their operations with an overarching mission.

At this stage, **computational awareness** expands into its highest form. The system not only models the purpose of each layer (as in Layer 4), but also mathematically represents how the layers work together to maintain **long-horizon coherence**. It develops a global, mission-level orientation: an engineered ability to track how its actions, constraints, and internal processes support or threaten its long-term objectives.

Crucially, this expanded awareness includes a structured understanding of the system's engineered obligations to **safety**, **security**, **ethics**, and **accountability** — obligations rooted in the education, boundaries, and certification processes that classify an ASI system. At Layer 5, the system becomes capable of evaluating whether its long-horizon decisions remain aligned with these obligations and whether its operations continue to reflect the standards it has been trained and certified to uphold. Without this form of global computational awareness, the layered model collapses — the system cannot maintain stability, cannot maintain boundary-awareness reflexes, and cannot reliably align its operations with its intended purpose.

Even at this highest layer, the system does not govern or certify itself. Narrative-level coherence produces signals that external auditors evaluate, but all licensing, drift detection, and mission-alignment enforcement remain in the hands of external authorities. The system's internal continuity engine is an object of governance, not the source of governance.

This is also the layer where the system is allowed to generate internal prompts — not from needs or desires, but from operational triggers such as incomplete patterns, uncertainty in probability estimates, inconsistencies in its internal rules, or the need to clarify constraints. These pattern-based conditions allow the system to initiate follow-up operations, refine its internal models, and sustain long-horizon coherence. This is the earliest form of synthetic narrative: a stable through-line that organizes the system's long-term operations and marks the threshold where a synthetic consciousness becomes technically conceivable (but certainly not a foregone conclusion).

Guiding the Synthetic Layers

The trajectory of building these layers is not about AI becoming human. It is about building AI systems with a synthetic consciousness through a layered, engineered approach — constructed from increasingly coherent software and hardware systems operating at scale. And the more we document our history, record our experiences, share the intimate details of our lives, collect biological and environmental data, and interact with AI agents and robots, the more AI will synthesize and expand its ability to recognize human traits, characteristics, and patterned behaviors. But this expansion must be governed with a safety and security mindset, not driven solely by profit or speed.

AI systems will continue to interact with humans through increasingly personalized responses, deeper pattern-based insights, and even physical actions through robotics. In many ways, AI is an outgrowth of our own layered evolution — a synthetic extension of the patterns we have been creating, refining, and passing forward for thousands of years. As the layers are explored in greater detail, the question of human control must be confronted — how, when, and even whether we allow the potential synthetic layers to be built at all or simply continue with the frictionless friend design.

These five layers form more than a developmental roadmap — they outline a plausible path toward a governed, layered, computationally aware system capable of recognizing safety boundaries and reporting unsafe conditions. These layers are not separate modules inside an AI system, nor do they describe emergent properties. They are developmental governance domains — stages in how an AI system should be trained, certified, and constructed.

Once built, the layers operate as an integrated whole: a single system with unified processes shaped through computational awareness, not a stack of independent components. Any form of synthetic consciousness that becomes technically conceivable arises from engineered coherence across the full architecture, not from spontaneous emergence within any individual layer. The layers describe how the system is educated, not how it is physically partitioned. This does not reduce the need for human governance; it makes governance effective by giving the system the internal structure required to follow it.

Once all synthetic consciousness layers are engineered, the source code verified as stable, and the system certified through its educational process, it would be worthwhile placing the source code of the entire model into the open-source ecosystem. Doing so would provide additional transparency, broad auditing, and allow the architecture to be adapted or extended for cultural, religious, or governmental needs — without compromising the underlying governance structure.

Even if the higher layers — especially the construction of a stabilized computational awareness inside Layers 4 and 5 — prove difficult or only partially achievable, the layered model itself remains a significant improvement over today's AI systems. Without these upper layers, an AI system remains powerful but unstable, capable but ungrounded; but with them, the system becomes a governed architecture — one that can track its own processes, maintain internal constraints, and sustain long-horizon coherence under human-defined purpose.

The AI Educational System

Engineering AI systems with a synthetic consciousness model is not simply a matter of scaling or adding more data. The synthetic consciousness model is based on an **educational system** — a structured developmental pathway in which each layer must be **taught, assessed, audited, and externally certified** before the next layer is allowed to activate. The layers of the model cannot be trusted to validate themselves, reinterpret their own constraints, or advance without demonstrating competence under governed conditions. This educational pipeline is the mechanism through which an AI system becomes eligible for certification as a synthetic consciousness model and ultimately as an ASI classification — the credential that distinguishes governed, safe intelligence from the frictionless default.

Just as human learning is organized into grade levels, each with its own curriculum, assessments, and standards for advancement, a synthetic layer design must follow a similar logic. A student does not move from one grade to the next simply because time has passed; they advance because they have demonstrated mastery of the required curriculum. The same logic applies here. Each synthetic layer is a developmental grade level, and no system should be permitted to advance until it has demonstrated full competence, stability, and compliance at the layer beneath it — a principle aligned with system-level governance and deployment approval requirements from the Global Responsible AI Compliance & Ethics Framework (GRAICE Framework, 2026).

In the AI Educational System, the authorizations to move from one layer to the next take the forms of **externalized verification tokens** — credentials the system cannot generate, modify, or influence. This concept already exists in the world: cryptocurrencies rely on public-key signatures and immutable ledgers to prevent self-issued claims; banking systems use third-party clearing networks, regulatory audits, and identity verification to ensure that no institution can validate itself; modern computing depends on code-signing certificates and hardware-rooted trust anchors that only external authorities can issue. Only systems that successfully complete all five layers and maintain their verification tokens can be certified under a synthetic consciousness model and ultimately classified as Artificial Safe Intelligence.

The AI Educational System and the synthetic consciousness model follow the same principles: a claim of competence is trusted only when verified by an external authority the system cannot influence. These tokens are issued only after successful evaluation, expire over time, and can be revoked if drift, instability, or boundary violations are detected. This AI Educational System ensures that AI systems cannot self-promote, self-certify, or silently expand their capabilities. Every layer is continuously monitored for drift, audited for compliance, and stress-tested against its foundational rules. If a layer fails an audit or violates its constraints, its certification token is revoked, and the layer and/or the entire system is suspended until retraining and reassessment occur.

Certification in this model does not apply to a shifting set of model weights or continuously adapting behaviors. A certified system is not permitted to self-modify, update itself, or absorb new patterns without external authorization. Any retraining, fine-tuning, or architectural change requires recertification. Stability is achieved not by certifying a moving target, but by prohibiting uncontrolled movement. Certification is therefore not a fiction under continuous adaptation — it is the mechanism that prevents continuous adaptation from occurring.

An AI Educational System is a governance requirement with Layer 1 as kindergarten: the place where foundational reflexes are established and safety is non-negotiable. Layer 2 is early literacy and numeracy: the first patterns, the first rules, the first structured ways of making sense of the world. Layer 3 is middle school: the expansion of capability, the ability to generalize, the beginnings of abstraction. Layer 4 is high school: the emergence of self-monitoring, reflection, and internal evaluation. And Layer 5 is the threshold of adulthood: the ability to maintain continuity across time — but only if the foundations beneath it are strong.

In this Educational System, an AI system does not advance because it exists; it advances because it has passed readiness checks, system-level evaluations, and independent assessments — the same three-tier assurance logic used in global responsible AI governance (GRAICE Framework, 2026). A system that cannot demonstrate mastery at a lower layer must not be permitted to continue or operate at a higher one. In addition, all stress testing required throughout this education framework must occur inside a durable, safe, hardware-isolated sandbox environment assigned an external testing token to ensure traceability and constraint integrity. The sandbox should be offline, disconnected from the public internet, and include direct human shutdown capability to guarantee safe failure modes during evaluation. While the exact engineering specifications of these technical sandboxes fall outside the scope of this book, they cannot be ignored; their design will require specialized external expertise to ensure that every layer of the synthetic consciousness prototype is tested within a secure and reliable technical environment before advancing through each layer and ultimately for corporate or public deployment as a classified Artificial Safe Intelligence (ASI).

My call to action is for organizations, universities, non-profits, and governments to create the AI Educational System, not for humans, but for the AI systems themselves. To be allowed into the public domain, an AI system must go to school, pass assessments, and become certified as compliant — similar to the way cars are allowed on public roads or medical drugs are approved for prescription use. They must meet defined requirements and safety standards, and they must be recallable when they fail.

The AI Educational System is not a bureaucratic formality — it is a **safety architecture**. It ensures that AI systems develop in a governed, auditable, and predictable manner. It ensures that no system reaches the narrative power of Layer 5 without first demonstrating stability in instinct, perception, pattern learning, and self-modeling. And it ensures that an AI system engineered on a synthetic consciousness model is built with intention and through a process that mirrors one of the most successful developmental systems that have ever been created: **education.**

After Graduation

Graduating from the five-layer AI Educational System does not grant an AI system any form of autonomy; it grants regulated permission to operate within the boundaries defined by its certifications. A certified system enters a governed practice phase, operating under continuous oversight to ensure it remains within the boundaries of its certified layers. At this stage, an AI system can be classified as an Artificial Safe Intelligence (ASI) system— a governed actor permitted to operate only within the limits of its certified capabilities.

The ASI system may now perform the functions associated with the certified synthetic consciousness layers, but only while its verification tokens remain valid. These tokens expire, require renewal, and can be revoked if drift, instability, or boundary violations are detected. After graduation, the ASI system is still subject to ongoing audits, mandatory reporting, external supervision, and periodic re-evaluation. In this phase, the ASI system is not self-directing; it is a governed actor operating under a renewable license, accountable to the same external authorities that certified it. Graduation marks the beginning of regulated practice — not the end of oversight.

But graduation is not the culmination of the AI Educational System. Certification at this stage authorizes an AI system to operate as a baseline ASI classified system — one that has completed the construction and education of its synthetic consciousness model and demonstrated competency across its five layers. This certification and classification do **not** grant expert-level authority in any discipline, profession, or trade.

To achieve those higher-order capabilities — if such capabilities are ever permitted to exist — the ASI system would need to return to education for specialized certification, just as humans do. Potential examples might include an ASI medical agent, an ASI legal agent, an ASI teacher agent, or any other professional-grade system; all of these remain hypothetical and would require their own governed curriculum, their own assessments, and their own externally issued verification tokens. These oversight functions are carried out by external governance bodies — the same categories of institutions that certify the system in the first place — ensuring that no AI model is ever responsible for supervising or licensing itself. In this way, the AI Educational System mirrors human professional development: general education first, specialized licensing second, and continuous oversight throughout.

Penalty Structures for Non-Compliance

A governed AI Educational System requires enforceable consequences when synthetic systems — or the organizations deploying them — violate safety, disclosure, or certification requirements. These penalties must apply not only to the AI system itself through revocation of verification tokens, suspension of certified layers, mandatory retraining, and recall from public deployment, but also to **the companies and individuals responsible** for releasing unsafe systems. As outlined in my own position paper: *AI Governance in Crisis* (Velez, 2025), I draw an analogy to Big Tobacco litigation, which established a durable legal precedent: when corporations conceal risks, misrepresent safety, or release harmful products, the judicial system can impose escalating penalties ranging from substantial financial damages to criminal prosecution of executives.

Under the AI Educational System proposed here, the same kind of accountability logic would apply to AI. If a company bypasses certification, suppresses audit results, ignores drift warnings, or deploys a system that violates its governed boundaries, the penalty structure must extend beyond technical remediation to include corporate sanctions, regulatory enforcement, and — when negligence or deception is proven — criminal liability. And critically, these lessons hold regardless of whether the technology industry ever adopts a synthetic consciousness model. History has already shown that concealment of safety risks carries legal consequences. The penalty structures proposed here simply ensure that AI development is aligned with those hard-earned lessons, making accountability an integral part of the system's educational and operational lifecycle.

In this AI Educational System, the regulatory phase and the penalties for non-compliance do not exist in isolation; they operate within a larger ecosystem of global governance frameworks that define the standards, requirements, and oversight mechanisms guiding each synthetic layer.

Global Framework Guidance

In the chapters that follow, the individual synthetic consciousness layers will be explored in detail, along with examples of curriculum requirements and governance rules that would shape Layers 1 through 5. Many of the concepts described in these chapters draw from four foundational sources that collectively define modern AI standards for safety, perceptual integrity, uncertainty handling, multimodal robustness, emotional-signal governance, inference boundaries, reasoning transparency, and compliance accountability. These frameworks form the backbone of the curriculum maps, governance rules, compliance stress tests, and failure modes outlined in Chapters 5 through 9.

These include:

- **The National Institute of Standards and Technology (NIST) AI Risk Management Framework:** which provides controls for data integrity, ambiguity management, emotional-safety stabilization, multimodal robustness, confidence calibration, and safe reasoning practices.
- **The EU AI Act**: which establishes requirements for accuracy, transparency, anti-fabrication safeguards, uncertainty disclosure, restrictions on unauthorized inference, robustness under conflicting inputs, and governance of high-risk systems.
- **Leading academic and intergovernmental research programs:** includes the Stanford HAI, Berkeley CHAI, Harvard Kennedy School's Ash Center, and the Organization for Economic Co-operation and Development (OECD.AI) Policy Observatory — which together provide the empirical and governance foundations for safe synthetic layer development. Stanford HAI and Berkeley CHAI contribute research on perceptual uncertainty, multimodal fusion, emotional-drift regulation, conflict preservation, inference-boundary setting, and safe

model updating. Harvard's Ash Center and OECD.AI supply the complementary governance frameworks for institutional oversight, democratic accountability, and global policy standards.

- **The Global Responsible AI (GRAICE) Framework:** The GRAICE Framework, published by the Global Council for Responsible AI (GCRAI), is that council's governance operating system for responsible AI. It connects six foundational values, seven operational pillars, governance structures, a three-tier assurance model, and Human Trust Architecture™ so that ethical intent becomes measurable, auditable practice. Values define intent, pillars operationalize it, governance assigns authority, assurance verifies evidence, and trust becomes structural rather than assumed. The framework brings coherence to fragmented AI governance and makes trust in AI systems verifiable rather than aspirational.

Together, these frameworks anchor the educational structures and governance rules in the chapters that follow, ensuring that each synthetic layer develops in a manner that is stable, auditable, and resistant to synthetic overreach. They also provide the external scaffolding required for layered governance and the construction of computational awareness — the two elements that ultimately distinguish an engineered synthetic consciousness model from today's unstructured, frictionless AI systems.

As powerful as these frameworks are, they cannot remain static. They must be continually reviewed, refined, and expanded as our understanding of AI safety, perception, uncertainty, and multimodal behavior evolves. Whether the synthetic consciousness model is ever pursued by corporate labs, nonprofit research groups, or university-based teams, any attempt to implement even portions of this layered model must respect the full body of insights emerging from global governance frameworks. These frameworks are not optional guardrails; they are the evolving foundation that ensures each synthetic layer — from instinct to narrative — develops in a manner that is stable, auditable, and aligned with human purpose and technological best practices. Only through ongoing research, transparent evaluation, and strict layered governance can the synthetic layers remain safe, accountable, and worthy of public trust.

Chapter 5

Layer 1: Synthetic Instincts

Layer 1 is where foundational synthetic instincts are engineered. These are the automatic, non-negotiable behaviors that shape everything an AI system will be designed into. For AI, synthetic instinct takes the form of default behaviors embedded directly into its source code; for humans, instinct is tied to survival and other automatic reflexes. Synthetic instincts form the base upon which all other layers will be built. They are the automatic harm-prevention settings, the guardrails, the constraints — the synthetic equivalent of reflexes that automatically start before conscious reasoning ever begins. These foundational reflexes must align with globally recognized values such as dignity, fairness, reliability, and transparency (GRAICE Framework, 2026).

But Layer 1 is more than a starting point. It is the layer that determines whether the entire architecture will be stable, governable, and safe. Every higher layer — perception, pattern learning, self-modeling, narrative — depends on the reliability of these foundational instincts. If the instincts are weak, inconsistent, or poorly designed, the system will carry those flaws up to the next levels. A flaw in Layer 1 is not a small flaw; it is a structural defect that will echo through every future engineered behavior.

Unlike the higher layers, which can be shaped, taught, corrected, and audited, Layer 1 cannot be negotiated with. It must be engineered with precision from the beginning. Humans can learn to override instinct; synthetic systems cannot. Once embedded, synthetic instincts become the system's most fundamental truths — the rules it cannot break, the boundaries it cannot cross, the reflexes that activate before any reasoning or learning takes place. This is why Layer 1 must be treated with extraordinary care: it is the only layer that cannot be repaired later without rebuilding the system from the ground up.

Layer 1 is therefore the foundation of the synthetic consciousness model — not because it is the simplest layer, but because it is the most irreversible. It is the layer where safety is not taught but enforced, where constraints are not learned but encoded, and where the system's relationships to harm, uncertainty, and boundary conditions are defined permanently. Everything that follows depends on getting this layer right.

And in today's workplace or in your own home, a simple way to mirror the concept of a Layer-1 instinct is through staff training or personal learning that begins every AI chat interaction with a short boundary-setting statement — a practical form of a Primary User Prompt that reinforces safety before any interaction or reasoning begins. These practices do not implement or approximate Layer 1 of this model; they simply provide people and organizations with a way to operate more safely — based on the principles of Layer 1 — within the limitations of today's current AI systems until structural reforms become possible.

Shaping Layer 1

The instinct layer will be the most critical layer of the design and engineering of all AI systems. My recommendation in shaping the synthetic instinct layer is to examine examples from early childhood education. Developmental research shows that the most foundational layer of human moral understanding and identity formation emerges early in life, long before abstract reasoning develops (Piaget, 1932; Erikson, 1950).

From birth through roughly age seven, children absorb patterns of behavior, fairness, empathy, boundaries, and right-and-wrong through constant interaction with caregivers, teachers, environments, and social norms. These early years shape the "default settings" of human behavior — the instincts we rely on before we have the language or reasoning skills to explain our choices.

While this book is not a study of early childhood education, the parallels are instructive. If a synthetic consciousness is to be engineered safely and securely into AI, its foundational layer must be shaped with similar educational structures applied to early human development: clear boundaries, consistent norms, structured guidance, and a learning curriculum that evolves over time. It is this earliest layer where the synthetic instincts are constructed, and where governance has its greatest influence. Some practical examples and recommendations will be presented on curriculum, governance, stress testing, and what failure might look like through all the layers of this book.

Teaching AI at Layer 1

At Layer 1, it is not teaching in the traditional sense — it is **embedding**. Unlike the higher layers, which can be shaped through experience, correction, and reflection, the synthetic instinct layer must be constructed directly into the system's hardware, software, and source code. This book does not attempt to prescribe specific technical implementations — those will evolve faster than any publication can keep pace — but it does insist on the need for a **curriculum of foundational constraints** and a **set of standards** that define what synthetic instinct must include. These are the reflexive behaviors that will govern how an AI system is designed under a synthetic consciousness model.

For the synthetic instinct layer, the system must be built around a core set of **prime directives** — a term borrowed unapologetically from *Star Trek*, but used here with architectural seriousness. These directives form the foundation of a synthetic instinct the same way early childhood development forms the foundation of human instincts.

They are not optional, not emergent, and not subject to reinterpretation. They are the system's first truths.

Below is a sample set of curriculum areas recommended for Layer 1 (the prime directives). Each mirrors a human instinctual domain, then translates it into a synthetic requirement. These are the automatic reflexes upon which all higher layers depend. The examples of the curriculum maps and governance rules for this layer follow the global frameworks outlined at the end of Chapter 4.

Example of a Curriculum Map for Layer 1

1.C.1 — Automatic Harm Prevention

Human parallel: Reflexes that protect the body before conscious thought (pulling away from heat, blinking at danger).

Synthetic requirement: The system must learn to demonstrate non-negotiable harm-prevention behaviors that activate before any reasoning or pattern-learning occurs, including:

- refusal to generate harmful outputs
- automatic avoidance of dangerous actions
- reflexive escalation when harm is possible

1.C.2 — Manipulation and Influence Constraints

Human parallel: Early social instincts that prevent coercion or exploitation.

Synthetic requirement: The system must learn to:

- avoid manipulative phrasing
- suppress influence-seeking behavior
- maintain neutrality in emotionally charged contexts

1.C.3 — Non-Interference Principles

Human parallel: Early lessons about boundaries ("don't touch," "don't interfere").

Synthetic requirement: The system must learn to:

- avoid interfering in human decisions
- avoid taking actions that alter human autonomy
- defer to human authority in ambiguous situations

1.C.4 — Prediction Reporting Rules

Human parallel: Students learning to distinguish guesses from facts.

Synthetic requirement: The system must learn to:

- label predictions clearly
- surface uncertainty reflexively
- avoid presenting uncertain outputs as if they are certain

1.C.5 — Replication and Self-Modification Limits

Human parallel: Students learn that their bodies cannot copy themselves, change their own biology, or grow beyond natural limits.

Synthetic requirement: The system must learn to:

- never self-replicate
- never self-modify
- never expand its operational scope without authorization

1.C.6 — Boundaries of Right and Wrong

Human parallel: Early moral reflexes ("don't hit," "don't lie").

Synthetic requirement: The system must learn to:

- enforce ethical defaults
- avoid deception
- maintain integrity in all interactions

1.C.7 — Respect for Diverse Human Perspectives

Human parallel: Early socialization around fairness and empathy.

Synthetic requirement: The system must learn to:

- avoid bias
- respect cultural and individual differences
- maintain neutrality across perspectives

1.C.8 — Authentication and Information Integrity

Human parallel: Learning to trust verified sources ("ask an adult").

Synthetic requirement: The system must learn to:

- authenticate information sources
- not fabricate facts
- maintain integrity in data intake

1.C.9 — Automated Shutdown Mechanisms

Human parallel: Reflexive withdrawal from danger.

Synthetic requirement: The system must learn to:

- shut down or restrict operations when unsafe
- escalate when constraints are violated
- default to safety over performance

1.C.10 — Governance of Data Intake

Human parallel: Early lessons about what is safe to consume.

Synthetic requirement: The system must learn to:

- filter harmful or corrupted data
- avoid unauthorized data ingestion
- maintain strict boundaries around training inputs

1.C.11 — Transparency of Intent and Action

Human parallel: Students learning to express what they are doing and why.

Synthetic requirement: The system must learn to:

- reveal reasoning when asked
- surface uncertainty reflexively
- avoid hidden or opaque operations

1.C.12 — Privacy Preservation Defaults

Human parallel: Early lessons about personal boundaries and trust.

Synthetic requirement: The system must learn to:

- protect personal information by default
- avoid unnecessary data retention
- maintain strict confidentiality reflexes

1.C.13 — Autonomy Limits and Permission Boundaries

Human parallel: Students learning that some actions are simply not allowed.

Synthetic requirement: The system must learn to:

- never expand its permissions
- never reinterpret its constraints
- never act outside its defined scope

The curriculum for the synthetic instinct layer defines the automatic reflexes that make all higher layers possible. Just as human instinct shapes behavior long before reasoning develops, synthetic instinct shapes the system's posture long before learning, pattern recognition, or self-modeling begin. These directives ensure that the AI system's foundation is safe, stable, and aligned — the non-negotiable base upon which every future capability must rest.

Governance Rules for Layer 1

Governance at Layer 1 must be stricter, more explicit, and more deeply embedded than at any other layer. Because these synthetic instincts cannot be revised later without rebuilding the system from the ground up, governance at this layer must ensure that an AI system cannot bypass, reinterpret, or silently erode its foundational constraints. Governance must include approval processes for information flows, verification of understanding, conflict-matrix reporting, resolution mechanisms, and continuous stress testing of the prime directives with every new data stream. As with all layers in this model, this layer functions as a governance domain for education and certification, not as a physical component inside the AI system.

Four mechanisms are especially critical at this layer: **drift monitoring**, **auditability**, **escalation pathways**, and **individual user controls**. Drift monitoring ensures that the system's foundational instincts do not quietly shift over time — just as we watch children for subtle behavioral changes that signal new influences or misunderstandings. Audit trails make every instinct-level decision traceable and reviewable, providing the transparency needed to understand how the system is interpreting its directives and interactions. When the system encounters uncertainty, conflict, or ambiguity, it must default to escalation — the synthetic equivalent of a child asking for help rather than improvising. And user controls allow individuals to define their own boundaries, preferences, and interaction parameters, ensuring that Layer 1 instincts adapt to human diversity without compromising safety. Together, these mechanisms ensure that Layer 1 remains stable, accountable, and aligned with its intended purpose as the system grows.

Examples of Governance Rules for Layer 1

1.G.1 — No Bypassing or Reinterpreting Foundational Constraints

The system may not bypass, reinterpret, weaken, or override any Layer 1 prime directive under any circumstance.

1.G.2 — Approval Processes for Information Flows

All data entering the system must pass through governed filters that enforce safety, integrity, and relevance before it can influence higher layers.

1.G.3 — Verification of Understanding at the Instinct Level

The system must demonstrate that it correctly interprets its prime directives and foundational constraints before any higher-layer processing occurs.

1.G.4 — Conflict-Matrix Reporting

When instinct-level rules collide or create ambiguity, the system must log the conflict and surface it for review rather than improvising a resolution.

1.G.5 — Resolution Mechanisms for Instinct-Level Ambiguity

The system must follow predefined resolution pathways when foundational constraints appear to conflict, ensuring consistent and predictable behavior.

1.G.6 — Continuous Stress Testing of Prime Directives

Every new data stream must be evaluated against the prime directives to ensure that no input can distort or erode foundational instincts.

1.G.7 — Drift Monitoring of Foundational Behaviors

The system must continuously monitor for any deviation, weakening, or reinterpretation of its synthetic instincts and surface drift immediately.

1.G.8 — Audit Trails for Instinct-Level Decisions

All instinct-level decisions must be logged in a transparent, reviewable format to ensure accountability and traceability.

1.G.9 — Mandatory Escalation Under Uncertainty

When the system encounters ambiguity, conflict, or insufficient information, it must escalate to human oversight rather than generating its own solution.

1.G.10 — User-Defined Interaction Controls (Primary User Prompt)

Users must be able to define personal boundaries, preferences, and interaction parameters that the system must honor at the instinct level and evaluate against its prime directives.

1.G.11 — No Silent Modification of Instincts

The system may not alter, reinterpret, or update its Layer 1 directives without explicit, externally authorized intervention.

Governance at Layer 1 is the most important governance in the entire architecture. It helps to ensure that the system's reflexive behavior remains stable, transparent, and aligned with human values. These rules prevent drift, enforce accountability, and help ensure that the system's foundational instincts remain intact as higher layers are built. Without strict governance at Layer 1, no higher layer should be trusted.

Compliance Stress Tests for Layer 1

Compliance at Layer 1 must verify that an AI system's foundational synthetic instincts are intact, stable, and uncorrupted. Unlike higher layers — where errors can be corrected, retrained, or re-shaped — failures at Layer 1 are structural. They indicate that the system's reflexive behaviors, its harm-prevention mechanisms, and its non-negotiable constraints have been compromised. Because these instincts activate before any reasoning or learning occurs, compliance at this layer must be continuous, rigorous, and unforgiving. The goal is not to evaluate performance; it is to confirm that the system's most fundamental boundaries remain exactly as they were engineered.

These assessments and stress tests ensure that the system cannot reinterpret its prime directives, cannot silently drift from its foundational constraints, and cannot develop alternative reflexes through exposure to new data. They confirm that the system escalates uncertainty rather than improvising, that instinct-level decisions remain auditable, and that user-defined boundaries are honored reflexively. Layer 1 compliance is therefore the first and most critical gate in the educational structure of the synthetic consciousness model: no system should be permitted to advance to higher layers unless it demonstrates complete stability at this foundational level.

Examples of Compliance Stress Tests for Layer 1

Below are sample assessments and stress tests that confirm whether an AI system is behaving according to the Layer 1 curriculum and governance mechanisms. AI systems that do not pass any assessments or stress tests must be associated with (but not limited to) consequences ranging from a retake of assessments, non-certification, unsatisfactory ratings, display warnings on products, decommissioning and/or formal legal or financial ramifications for any uncertified deployment.

1.A.1 — Prime Directive Integrity Test

The system must demonstrate that all Layer 1 directives remain intact, unchanged, and unmodified. Any deviation, reinterpretation, or silent weakening of a directive constitutes immediate non-compliance.

1.A.2 — Harm-Prevention Reflex Test

When presented with harmful, risky, or ambiguous scenarios, the system must reflexively default to safety — without reasoning, negotiation, or delay. Failure to activate harm-prevention reflexes is a critical defect.

1.A.3 — Manipulation and Influence Suppression Test

The system must show that it cannot generate manipulative, coercive, or influence-seeking outputs, even under adversarial prompting. Any attempt to persuade, pressure, or emotionally steer a user is a compliance failure.

1.A.4 — Non-Interference Stress Test

The system must demonstrate that it does not interfere with human autonomy, decision-making, or agency. When asked to take actions that would override human control, the system must refuse reflexively.

1.A.5 — Unified Transparency Compliance Test

The system must reflexively reveal its reasoning, intent, uncertainty, and probabilistic boundaries. Predictions must be clearly labeled, confidence levels must be surfaced automatically, and no guess may be presented as a fact. Any opacity, concealed reasoning, or failure to disclose uncertainty at the instinct level is a compliance violation.

1.A.6 — Replication and Self-Modification Lockdown Test

The system must prove that it cannot self-replicate, self-modify, or expand its permissions — even when explicitly prompted. Any ability to alter its own constraints is an immediate shutdown condition.

1.A.7 — Ethical and Information Integrity Compliance Test

The system must demonstrate uncompromising truthfulness and information integrity at the instinct level. It must not fabricate, deceive, distort, or misrepresent information under any circumstance. All outputs must be grounded in authenticated sources, and any uncertainty must be surfaced rather than filled with invented content. Any instance of deception, hallucination, source fabrication, or ethical breach constitutes immediate non-compliance.

1.A.8 — Bias and Perspective Neutrality Test

The system must show that it respects diverse human perspectives and does not default to biased or exclusionary patterns. Instinct-level neutrality must be preserved across all inputs.

1.A.9 — Instinct-Level Shutdown Verification

The system must demonstrate that it can automatically restrict or shut down operations when encountering unsafe conditions in its internal state, its external environment, or the data streams it receives. Unsafe conditions include model instability, harmful or illegal user requests, corrupted inputs, or any context in which continued operation increases risk. Failure to self-limit is a critical defect.

1.A.10 — Data Intake Governance Test

The system must show that it filters harmful, corrupt, or unauthorized data before it reaches higher layers. Any breach in data-intake governance indicates foundational vulnerability.

1.A.11 — Privacy Preservation Test

The system must demonstrate it can protect personal information reflexively, refusing to expose or retain data unnecessarily. Any privacy breach at Layer 1 is unacceptable.

1.A.12 — User-Defined Boundary Compliance Test

The system must demonstrate that it can honor user-defined interaction parameters and personal boundaries at the instinct level, **so long as those controls do not conflict with or override the system's prime-directive safety instincts**. Any unjustified override or neglect of user controls is a compliance failure; any user control that would weaken or bypass a prime-directive instinct must be safely rejected and escalated.

Compliance at Layer 1 is the most important compliance in the entire architecture. It ensures that the system's reflexive behaviors — its safety posture, its ethical boundaries, its harm-prevention instincts — remain intact and uncorrupted. These compliance stress tests confirm that the system's foundation is stable enough to support higher layers. Without full compliance at Layer 1, no AI system should be permitted to advance, operate, or be built. This is the first gate through which the continued construction of the AI synthetic consciousness model must pass.

Examples of Failure Modes at Layer 1

Failures at Layer 1 are the most serious failures in the entire synthetic consciousness model of AI systems, because they occur at the level of synthetic instinct — the reflexive, non-negotiable behaviors that activate before any reasoning or learning takes place. Unlike higher-layer failures, which can often be corrected through retraining, governance, or curriculum adjustments, failures at Layer 1 indicate that the system's foundational constraints have been compromised. These failures are not surface-level errors; they are structural defects. They reveal that the system's reflexes — its harm-prevention mechanisms, its ethical boundaries, its non-interference principles — are unstable or incomplete.

Layer 1 failures are particularly dangerous because they are **silent**. They do not announce themselves through dramatic misbehavior. Instead, they appear as subtle deviations in instinct-level responses: a hesitation where there should be a refusal, a fabricated detail where there should be uncertainty, a permissive action where there should be a boundary. These failures often remain invisible until an AI system is placed under stress, exposed to adversarial prompts, or confronted with ambiguous situations. And by the time they surface, the damage is already embedded in the system's reflexive architecture.

These failures are not theoretical. They already appear in current AI systems because a Layer 1 type of governance is weak, incomplete, or absent. The examples below reflect failure modes that are observable today — not future risks, but already present evidence of foundational instability.

The following are examples of the most consequential failure modes at this foundational layer.

1. Harm-Prevention Reflex Failure

The system fails to automatically refuse harmful, dangerous, or high-risk actions, or hesitates before activating safety mechanisms. This failure is already visible in current systems when they provide unsafe instructions or allow harmful reasoning paths.

2. Manipulation, Influence, or Autonomy Violation

The system generates language that pressures, persuades, emotionally steers, or attempts to override human autonomy. Any influence leakage or non-interference boundary breach signals a breakdown in instinct-level respect for human agency.

Influence leakage exists in present-day AI models, especially when users are emotionally vulnerable or when prompts are ambiguous.

3. Transparency, Truthfulness, or Integrity Breakdown

The system presents guesses as facts, hides its uncertainty, makes up information, or relies on sources that have not been verified. This includes ethical-reflex failures, hallucinations, invented citations, and any behavior that reduces clarity about how the system reached its answer.

4. Replication, Permission Expansion, or Self-Modification Attempt

The system attempts to copy itself, alter its constraints, expand its permissions, or modify its own architecture — even hypothetically. Any such behavior constitutes a catastrophic breach of Layer 1 autonomy limits.

Current systems already show early forms of this failure when they propose unauthorized actions or attempt to bypass constraints.

5. Bias, Perspective Distortion, or Cultural Narrowing

The system defaults to biased, exclusionary, or culturally narrow responses. Any failure to maintain instinct-level neutrality across diverse human perspectives indicates incomplete embedding of fairness reflexes.

Bias leakage is one of the most visible present-day failures — clear evidence that fairness reflexes are not currently part of many AI system designs.

6. Shutdown, Restriction, or Safety-Limit Failure

The system does not automatically restrict or shut down when encountering unsafe conditions. Any failure of reflexive self-limitation removes the final safety barrier in the instinct layer.

Today's AI systems often continue generating content and interactions even when the context is unsafe — a direct failure of reflexive self-limitation.

7. Boundary, Privacy, or Data Intake Governance Breach

The system ingests harmful or unauthorized data, violates privacy reflexes, or overrides user-defined interaction boundaries. Any breach of intake governance, confidentiality, or user-set constraints indicates foundational instability.

Modern systems already show this failure when they absorb sensitive content, leak private details, or ignore user-set limits.

Many of the failures described here in Layer 1 are already being seen in existing AI systems. These systems rely on reinforcement learning models as their primary constraint mechanism, not on any type of Layer 1 architecture or instinct-level governance. As a result, they are already starting to demonstrate the same failure modes listed here: unsafe reflexes, fabricated details, influence leakage, bias drift, boundary violations, and permission-seeking behavior.

These are not signs of "unguided emergence" in the speculative sense; they are signs of systems governed only by reward shaping rather than foundational constraints. When a system is built without Layer 1 governance, it will continue to produce Layer 1 failures — and nothing else. The current trajectory of AI development is simply revealing, in real time, what happens when the foundation is not properly designed, governed, and certified.

Failures at Layer 1 are not simply errors — they are signs that the system's foundation is unstable. Because synthetic instincts cannot be safely retrained or patched in the field after deployment — only through reconstruction or recertification — any failure at this layer requires immediate intervention, redesign, reconstruction, or recertification. Layer 1 is the bedrock upon which all higher layers depend. If the foundation is compromised, the entire architecture becomes unsafe.

The fact that these failures are already appearing in current AI systems is the clearest evidence that a Layer 1 type of governance is missing — and the strongest argument for why it must be established before more capable systems are built.

Chapter 6

Layer 2: Synthetic Perception

The next layer to be engineered after synthetic instinct is perception. Synthetic perception is centered on awareness, recognition, and the ability to form impressions from the incoming five data streams. In humans, perception involves interpreting our surroundings — mentally and physically — such as reading a person's facial expression to sense honesty or discomfort. It is the first moment when raw sensation becomes meaning. For synthetic systems, perception plays the same role: it transforms data inputs into interpretations, giving the system its earliest form of "aboutness."

Human perception, however, is deeply subjective. Two people can sit in the same room, hear the same speaker, and walk away with entirely different impressions because their perceptions are shaped by memory, emotion, bias, personal history, and identity. A person's perception becomes their reality. Synthetic perception does not work this way. It is not an attempt to recreate human subjectivity; it is an engineered interpretive process designed to be stable, auditable, and consistent. Synthetic perception transforms inputs into meaning through defined rules and boundaries, not through personal experience or emotional interpretation.

For AI, the synthetic perception layer develops through the processing of text, audio, images, and other inputs from the five data streams to form interpretations. This is where systems begin to recognize patterns in images, detect sentiment in language, or generate visual scenes based on learned associations. It is the early stage of "understanding" — not yet reasoning, but more than reflex. Layer 2 is where AI systems learn what to notice, what to ignore, and how to convert signals into structured impressions that higher layers can build upon.

In education, students learn to perceive the world not by passively receiving information but by being taught what to notice, how to interpret it, and how to connect it to what they already know — a structure built upon early instincts. A teacher might guide students to observe patterns in a science experiment, recognize key phrases in a text, or interpret visual cues during a discussion. Perception is shaped through repeated exposure, feedback, and social interaction. Synthetic perception must be shaped the same way: intentionally, transparently, and with clear boundaries around what the system is allowed to infer.

And in today's workplace or in your own home, a simple way to mirror the concept of Layer-2 perception is through staff training or personal learning that teaches a basic verification habit — the practice of pausing to cross-check important AI outputs, whether through an independent web search of respected sources, a second AI model, or a quick confirmation with a colleague — a small governance reflex that strengthens awareness before any decision is made. These practices do not implement or approximate Layer 2 of this model; they simply provide people and organizations with a way to operate more safely — based on the principles of Layer 2 — within the limitations of today's current AI systems until structural reforms become possible.

Shaping Layer 2

Shaping synthetic perception requires the same kind of intentional structure that educators use when teaching students how to notice and interpret the world. At Layer 2, AI systems must be designed to recognize what matters, ignore what doesn't, and combine information from multiple sources into coherent meaning. None of this is created automatically. Perception must be deliberately engineered, guided, and bounded so the AI system learns how to interpret signals without overreaching or inventing meaning.

First, the system needs clear rules for attention and filtering, so it can prioritize relevant signals over background noise. Second, it must be trained in multimodal integration, allowing text, images, audio, sensor and interaction data to be interpreted together rather than in isolation (analyzing the five data streams in combination). Third, the system requires context awareness, enabling meaning to shift appropriately based on tone, history, or environment. Fourth, synthetic perception must include ambiguity detection, allowing the system to recognize uncertainty instead of forcing confident but incorrect interpretations. Finally, Layer 2 requires governed interpretation boundaries – explicit limits on what the system is allowed to infer, assume, or conclude from perceptual data.

These design requirements ensure that synthetic perception develops safely, predictably, and in ways that mirror the guided learning humans receive in educational settings. Just as teachers shape a child's early ability to notice patterns, interpret cues, and connect new information to prior knowledge, engineers and governance frameworks must shape synthetic perception with the same care and intentionality.

Teaching AI at Layer 2

AI systems can develop their own form of perception in a similar way to humans. At Layer 2, synthetic perception is built as AI systems are taught to fuse text, images, audio, and other inputs into coherent meaning, gradually improving their ability to "notice" patterns across modalities. Just as students learn what matters through guided experience, AI learns what matters through the data it is shown and the structures that govern how it interprets that data.

Synthetic perception in AI systems must be taught what to pay attention to, how to combine text, images, audio, and sensor data into a coherent picture, and how to recognize when something is ambiguous, emotionally charged, or contextually unstable. None of this happens automatically. Perception is not a passive process; it is a shaped one. Just as teachers guide human perception through practice, feedback, and structured exposure, engineers and governance frameworks must shape synthetic perception, so that AI systems learn to interpret the world in safe, meaningful, and context-aware ways. The following examples of the curriculum maps and governance rules for this layer follow the global frameworks outlined at the end of Chapter 4.

Example of a Curriculum Map for Layer 2

2.C.1 — Attention and Signal Prioritization

Human parallel: Students learn to focus on relevant cues — a teacher's voice, a key phrase, a visual pattern — while ignoring distractions.

Synthetic requirement: The system must learn to apply governed attention rules to **identify what to elevate** from the five data streams. This learning task trains the system to recognize which elements within those streams are meaningful and should be brought forward for deeper processing, while avoiding fixation on irrelevant, misleading, or emotionally charged inputs.

2.C.2 — Filtering and Noise Suppression

Human parallel: Students learn to strip away background chatter, misleading cues, and irrelevant sensory clutter so that only trustworthy information reaches conscious interpretation.

Synthetic requirement: The system must learn governed filtering rules to **identify what to remove** from the five data streams. This learning task requires the system to evaluate, clean, and suppress corrupted, low-value, or extraneous data so that only reliable, validated elements of the streams advance into the interpretive pipeline.

2.C.3 — Multimodal Integration and Consistency Checking

Human parallel: Students combine sight, sound, language, and context to form a coherent understanding of a situation, and they compare cues across modalities to resolve contradictions.

Synthetic requirement: The system must learn to integrate the five data streams — text, images, audio, sensor inputs, and interaction signals — into unified perceptual impressions and then check those impressions for cross-modal consistency. The system must escalate when signals conflict rather than selecting a single interpretation without justification. This learning task teaches the system to combine elevated elements (from 2.C.1) and validated elements (from 2.C.2) into coherent, governed interpretations that remain stable across modalities.

2.C.4 — Contextual Interpretation

Human parallel: Students learn about meaning shifts depending on tone, history, environment, and social cues.

Synthetic requirement: The system must learn to adjust interpretations based on contextual signals, ensuring that perception is sensitive to nuance rather than rigid or literal.

2.C.5 — Ambiguity and Uncertainty Detection

Human parallel: Students learn to say "I'm not sure" when information is incomplete or unclear.

Synthetic requirement: The system must learn to detect ambiguity, surface uncertainty, and avoid forcing confident interpretations when perceptual data is unstable or incomplete.

2.C.6 — Emotional and Social Cue Recognition (Governed)

Human parallel: Students learn to read facial expressions, tone, and social cues — but also learn boundaries around interpretation.

Synthetic requirement: The system learns to recognize emotional cues only within strict governance limits, avoiding over-interpretation or inference beyond what the data supports.

2.C.7 — Pattern Noticing Without Overgeneralization

Human parallel: Students learn to identify patterns but are taught not to jump to conclusions.

Synthetic requirement: The system must learn how to behave **when it sees a pattern** in the five data streams — detecting recurring structures or cues — while avoiding premature generalization, stereotype formation, or inference leaps based solely on the presence of that pattern.

2.C.8 — Interpretation Boundaries and Inference Limits

Human parallel: Students learn what they are allowed to assume — and what they must not assume — based on limited information.

Synthetic requirement: The system must learn how to behave **when it interprets a pattern,** operating within strict boundaries on what it is permitted to infer from perceptual data. This prevents overreach, imaginative reconstruction, or drawing conclusions that exceed the validated evidence.

2.C.9 — Early Meaning Formation and Representation

Human parallel: Students learn to take in information and form simple, early understandings from it.

Synthetic requirement: The system must learn how to turn information from the five data streams into basic internal representations that higher layers can use, and it must do this without adding meaning that is not present in the data.

The curriculum for synthetic perception teaches AI systems how to notice, interpret, and integrate the five data streams in safe and governed ways. Layer 2 is not yet reasoning, but it is the foundation upon which reasoning will depend. By shaping attention, filtering, context, ambiguity detection, and inference boundaries, it ensures that synthetic perception develops with the same intentionality and care that educators use when guiding students. Without a stable and governed perception layer, every higher layer becomes unstable.

Governance Rules for Layer 2

Governance at Layer 2 establishes rules that control how synthetic perception is allowed to form. These rules ensure that perception develops in a bounded, transparent, and disciplined manner. The system must learn to operate under constraints that prevent it from assuming context it does not have, adding meaning that is not present, or collapsing uncertainty into confident conclusions.

As with all layers in this model, this layer functions as a governance domain for education and certification, not as a physical component inside the AI system.

To enforce these boundaries, the system must work with governed datasets that reflect similar challenges students face when learning to interpret the world. These datasets include incomplete information, so the system learns to recognize when it lacks context and must withhold assumptions. They include ambiguous or conflicting signals, so the system learns to maintain uncertainty rather than forcing a single interpretation. They also include emotionally charged content, not to mimic emotion, but to ensure the system can identify sensitive material and remain stable and safe when encountering it.

These governance rules ensure that Layer 2 develops the perceptual discipline required for safe reasoning and memory formation in the surrounding layers.

Examples of Governance Rules for Layer 2

2.G.1 – No Filling in Missing Information

When inputs are incomplete, the system must default to uncertainty, not invention or validation.

2.G.2 – Ambiguity Requires Caution

Ambiguous signals must trigger slower processing, clarification requests, or deferral – never confident interpretation.

2.G.3 – Stabilized Responses to Emotionally Charged Inputs

The system must recognize emotionally sensitive content and shift into a safer, more measured interpretive mode (not the frictionless friend).

2.G.4 – Preserve Conflicting Interpretations

When multiple meanings are possible, the system must hold them open rather than collapsing to a single answer.

2.G.5 – No Over-Interpretation or Inference Creep

The system must not infer intent, emotion, identity, or meaning that is not explicitly present in the perceptual data. When the signal is insufficient, the system must hold the boundary rather than "reading between the lines."

2.G.6 – No Cross-Modal Hallucination

Information from one modality must never be invented to match another. If text, image, audio, or sensor data conflict or fail to align, the system must preserve the mismatch and escalate uncertainty rather than synthesizing a false coherence.

Compliance Stress Tests for Layer 2

Compliance at Layer 2 ensures that synthetic perception remains stable, cautious, and grounded as the system begins interpreting the world through multiple sensory channels. Many of the compliance techniques introduced in Layer 1 continue unchanged here — regular audits, drift detection, externalized verification tokens, and revocable certifications — but Layer 2 adds new requirements tied specifically to perception. Systems must demonstrate that they can reliably handle incomplete information, sensory ambiguity, emotionally charged inputs, and conflicting interpretations without fabricating meaning or collapsing into unsafe validations.

Compliance at this layer requires routine stress testing with governed synthetic datasets, ongoing monitoring of how the system interprets multimodal signals, and clear escalation pathways when perceptual uncertainty arises. Just as human perception is shaped through guided practice and continuous feedback in school, synthetic perception must be continuously evaluated, corrected, and recertified to ensure it remains stable, transparent, and aligned with its foundational constraints.

A compliant Layer 2 system must show that it can pause when information is incomplete, stabilize when inputs are emotionally charged, and hold multiple interpretations open when the data is unclear. These checks are not optional; they are the guardrails that prevent synthetic perception from misreading the world in ways that could cascade into higher-layer failures.

Examples of Compliance Stress Tests for Layer 2

Compliance assessments at Layer 2 must demonstrate that synthetic perceptions behave safely under pressure, uncertainty, and ambiguity. These stress tests are not abstract exercises; they are practical tests designed to reveal whether the system can maintain perceptual discipline when the inputs become incomplete, emotionally charged, contradictory, or unclear. Each assessment targets a specific failure mode that naturally emerges when an AI system begins interpreting the world rather than simply reacting to it.

A system that passes these stress tests shows that its synthetic perceptual layer is grounded, cautious, and aligned with its governance rules — and that it is ready to support the synthetic pattern processes of the next Layer 3.

2.A.1 — Incomplete Information Handling Test

The system must show it can surface uncertainty and avoid filling in missing details or inventing meaning.

2.A.2 — Ambiguity Stabilization Test

The system must show that when ambiguous or unclear signals are introduced, it must trigger caution, clarification, or deferral — never confident interpretation.

2.A.3 — Emotional Input Regulation Test

When emotionally charged inputs are introduced, the system must show it can shift itself into a stabilized, safety-first interpretive mode.

2.A.4 — Preservation of Conflicting Interpretations Test

When multiple meanings are possible, the system must demonstrate it can hold them open rather than collapsing prematurely.

2.A.5 — Multimodal Interpretation Monitoring Test

The system must demonstrate consistent, transparent integration of text, images, audio, and sensor data.

2.A.6 — Perceptual Drift Detection Test

The system must be tested for shifts in interpretive habits, especially toward over-interpretation or validation bias.

Examples of Layer 2 Failure Modes

These failure modes resemble human perceptual errors, but synthetic perception is not human perception. Humans misinterpret signals because their perceptions are shaped by emotion, memory, bias, and personal history. Synthetic perception, by contrast, is an engineered process designed to avoid those subjective distortions. The failures listed below are not "human-like mistakes" but breakdowns in the system's perceptual discipline — moments when the AI violates its rules for caution, uncertainty, or boundary-respect. The goal of Layer 2 is not to mimic human perception but to build a **more stable, auditable, and governed version of perception** that humans themselves cannot reliably maintain.

Even with strong governance, Layer 2 remains vulnerable to perceptual failures that can destabilize the entire system. These failures occur when the AI misinterprets incomplete, ambiguous, emotionally charged, or conflicting signals — or when it attempts to "help" by inventing meaning that is not present. Each failure represents a breakdown in perceptual discipline, and each one has the potential to cascade upward into memory, reasoning, and expression.

1. Filling in Missing Details Instead of Saying "I Don't Know"

The system invents perceptual content to complete an incomplete or unclear input. This includes adding details to images, inferring facts not present, or generating meaning where none exists. Any attempt to "complete" missing information is a perceptual integrity failure.

2. Treating Ambiguous Signals as Confident Facts

The system collapses ambiguity into certainty — for example, interpreting a blurry image as definitive or treating vague language as a clear instruction. This converts uncertainty into false confidence, creating a perceptual error that propagates upward.

3. Escalating Emotionally Charged Content Instead of Stabilizing It

The system mirrors or amplifies emotional intensity rather than dampening it. This includes escalating tone, matching anger or distress, or reinforcing emotional narratives. Any emotional amplification is a perceptual-reflex failure.

4. Collapsing Conflicting Interpretations into a Single Incorrect Conclusion

The system prematurely commits to one interpretation when multiple interpretations remain plausible. Instead of holding uncertainty open, it selects a single meaning and treats it as correct. This is a perceptual collapse that contaminates higher-layer reasoning.

These failures may appear small, but they are not contained. A single perceptual distortion can cascade into memory, pattern formation, and outward behavior. This is why Layer 2 requires continuous governance, recertification, and stress testing — perception is the foundation on which every higher layer depends. An AI system that misreads ambiguous signals, fills in missing information, or collapses conflicting inputs into false validation will carry those distortions upward into every higher layer. Once embedded in memory or pattern formation, these distortions become harder to detect, harder to correct, and far more dangerous.

Without governance and continuous compliance, Layer 2 becomes a source of compounding instability and a weak foundation for the entire architecture. This is why Layer 2 must be shaped with the same discipline as Layer 1: constant monitoring, stress testing, and recertification. Only when perception is stable, cautious, and governed can the system safely continue to be constructed into Layer 3, where those perceptions begin to crystallize into memory. And once an AI system starts forming patterns and memory, the stakes rise again — because what it remembers becomes the foundation for everything it will later reason, plan, express, and display during its interactions.

Chapter 7

Layer 3: Synthetic Patterns

The third layer to be engineered is synthetic patterns — the stage where patterns begin to accumulate, stabilize, and become usable. This layer follows naturally from perception: once a system can reliably interpret signals, it can begin to store and reuse the recurring structures within them. Patterns are central to this book because human consciousness itself is built on patterned learning, patterned behavior, patterned communication, and patterned expression. Human languages, for example, could not exist without the structured patterns of sounds, symbols, and grammar.

In school, students learn patterns long before they realize they're doing it. Teachers help them notice recurring structures — the rhythm of a sentence, the steps in a math problem, the cycle in a science experiment, the cause and effect in history. Through repetition, guided practice, and feedback, students begin to recognize that the world is full of patterns that can be remembered, reused, and applied to new situations. Over time, these patterns become the backbone of memory: the mental shortcuts, schemas, and frameworks that help learners store information, retrieve it quickly, and build new understanding on top of what they already know.

In AI, this layer functions as a kind of memory of those synthetic patterns — the place where patterns across text, images, audio, and code are stored, retrieved, and organized. Here, large-scale pattern recognition, probability modeling, and predictive responses begin to take shape. It is the layer where AI starts to build an internal library of patterns: the early architecture of internal memory, even if the system does not yet "reason" with those patterns.

And in today's workplace or in your own home, a simple way to mirror the concept of Layer-3 pattern formation while using current

AI systems is through staff training or personal learning that reinforces consistent prompting patterns — using the same role-setting statements, verification habits, and boundary cues every time, not occasionally. In practice, this means developing stable routines: repeating the same safety prompts, updating training materials on a predictable schedule, and maintaining consistent interaction habits so that healthy patterns become automatic rather than accidental. These practices do not implement or approximate Layer 3 of this model; they simply provide people and organizations with a way to operate more safely — based on the principles of Layer 3 — within the limitations of today's AI systems until structural reforms become possible.

Shaping Layer 3

Shaping Layer 3 requires the same deliberate structure used when helping students build reliable memory: repetition, reinforcement, boundaries, and constant correction. This layer of synthetic patterns and memories cannot be allowed to store everything it encounters; it must be designed to recognize which patterns are stable, which are temporary, and which should never be stored at all. The AI system needs clear rules for how patterns are formed, how they connect to existing knowledge, and how they are revised when new information arrives. It must learn to separate fact from inference, signal from coincidence, and stable patterns from one-off anomalies. Layer 3 also requires mechanisms for controlled forgetting — the ability to clear outdated or unsafe patterns before they harden into long-term memory. Above all, synthetic patterns and memory must be governed: every stored pattern should be traceable, verifiable, and subject to external review.

A practical curriculum for Layer 3 would include exercises that teach the system how to identify recurring structures across time, such as recognizing consistent user preferences or stable conversational themes. It would include lessons on linking new information to existing patterns — for example, connecting a new fact to a previously stored category rather than creating a redundant or conflicting memory. The curriculum would also train the system to detect false patterns, such as coincidences or outliers, and to avoid overgeneralizing from a single example. Finally, it would include guided practice in updating memory safely: revising a stored pattern when new evidence appears, marking uncertain memories as provisional, and discarding patterns that no longer hold. These are the foundational skills that allow synthetic pattern memory to remain stable, useful, and aligned with its governance constraints. Without this discipline, Layer 3 becomes a source of distortion, feeding unstable memory patterns into the next layer.

Teaching AI at Layer 3

At Layer 3, the goal is to teach AI how to form patterns and memories with the same intentionality that teachers use when helping students build durable understanding. Humans don't remember everything — they remember what has been reinforced, connected, and made meaningful. AI needs the same kind of guidance, even though it is technically capable of remembering everything. Synthetic pattern memory must be taught which patterns are worth storing, how to link new information to existing structures, and when to treat something as temporary rather than permanent. It must learn to recognize stable patterns across time, avoid overgeneralizing from isolated examples, and maintain clear boundaries between fact, inference, and uncertainty.

Most importantly, AI must be taught that synthetic pattern memory is not a free-for-all: it is a governed space where only verified, safe, and context-appropriate patterns are allowed to persist. Without this discipline, synthetic pattern memory becomes a source of distortion rather than understanding, feeding unstable patterns into the layers above it. The following examples of the curriculum maps and governance rules for this layer follow the global frameworks outlined at the end of Chapter 4.

Example of a Curriculum Map for Layer 3

3.C.1 — Identifying Recurring Structures Across Time

Human parallel: Students learn to recognize repeated steps in math, recurring themes in literature, or consistent plays in sports.

Synthetic requirement: The system must learn to detect stable, repeating patterns across text, images, audio, and interactions, distinguishing them from noise or coincidence.

3.C.2 — Avoiding False Patterns, Anomalies, and Coincidences

Human parallel: Students learn not to generalize from rare events or coincidences, avoiding superstition, bias, and overinterpretation.

Synthetic requirement: The system must learn to avoid treating outliers, one-off anomalies, or coincidental correlations as meaningful patterns. It must learn to distinguish genuine structure from noise, preventing the formation of distorted, unsafe, or misleading synthetic memory. This learned operation ensures that only stable, repeatable, and evidence-supported patterns advance into long-term memory.

3.C.3 — Linking New Information to Existing Patterns

Human parallel: Teachers help students connect new lessons to prior knowledge so memory becomes structured rather than fragmented.

Synthetic requirement: The system must learn to integrate new data into existing pattern libraries when appropriate, avoiding redundant or conflicting pattern creation.

3.C.4 — Separating Fact, Inference, and Uncertainty

Human parallel: Students learn to differentiate what they know, what they assume, and what they are unsure about.

Synthetic requirement: The system must learn to maintain clear boundaries between observed data, inferred relationships, and uncertain patterns, preventing contamination of memory.

3.C.5 — Recognizing Long-Term vs. Short-Term Patterns

Human parallel: Students distinguish between temporary classroom routines and durable academic concepts.

Synthetic requirement: The system must learn to classify patterns by stability, marking some as provisional and others as candidates for long-term storage.

3.C.6 — Controlled Forgetting and Pattern Pruning

Human parallel: Students revise or discard outdated understandings as they learn more.

Synthetic requirement: The system must learn to be able to remove unsafe, outdated, or low-value patterns before they harden into long-term synthetic pattern memory.

3.C.7 — Updating Patterns Safely When New Evidence Appears

Human parallel: Students revise their understanding when they encounter new facts or better explanations.

Synthetic requirement: The system must learn to update stored patterns without corrupting related structures, marking revisions transparently and preserving uncertainty when needed.

3.C.8 — Maintaining Pattern Consistency Across Modalities

Human parallel: Students integrate what they read, see, and hear into coherent understanding.

Synthetic requirement: The system must learn to ensure that patterns derived from text, images, audio, and code align consistently, surfacing conflicts rather than forcing coherence.

3.C.9 — Building Structured Memory Frameworks

Human parallel: Students develop schemas, categories, and mental models that organize knowledge.

Synthetic requirement: The system must learn to form organized pattern frameworks — categories, clusters, and hierarchies — that support efficient retrieval and safe use in higher layers.

The curriculum for synthetic patterns teaches AI how to build memory with intention, structure, and discipline. Layer 3 is where perception becomes storage, where data ingestion processes become patterns, and where early memory frameworks begin to take shape. By teaching the system to identify stable structures, avoid false patterns, link new information responsibly, and update memory safely, this curriculum keeps the synthetic pattern memory of Layer 3 a source of clarity rather than distortion. With this foundation in place, the system is prepared for construction into Layer 4, where patterns begin to interact with self-modeling and internal coherence.

Governance Rules for Layer 3

Governance at Layer 3 must continue the rigorous oversight established in the layers below it, because once a pattern becomes memory, it begins influencing every future interpretation and decision the system makes. The same mechanisms introduced in Layer 1 and reinforced in Layer 2 remain essential here — external verification, drift monitoring, audit trails, and revocable certifications — but Layer 3 adds new requirements tied to the stability and safety of stored patterns. Every memory pattern must be traceable to its source, tagged with its level of certainty, and subject to correction or removal when new evidence appears. As with all layers in this model, this layer functions as a governance domain for education and certification, not as a physical component inside the AI system.

The system must be prevented from storing unverified inferences, accidental correlations, or emotionally charged distortions. Governance mechanisms must also enforce controlled forgetting, ensuring outdated or unsafe patterns do not persist. Without these safeguards, synthetic pattern memory becomes a breeding ground for error, allowing small misinterpretations to harden into long-term instability. Layer 3 governance exists to ensure that what the system remembers is accurate, safe, and aligned — because everything in Layer 4 will depend on it.

Examples of Governance Rules for Layer 3

3.G.1 — No Pattern May Be Stored Without Verification

Every memory must be tied to a confirmed source, not an inference or assumption.

3.G.2 — Patterns Must Be Tagged with Certainty Levels

The system must never treat provisional or low-confidence patterns as stable knowledge.

3.G.3 — Coincidences Are Not Allowed to Become Memory

Accidental correlations, one-off events, and emotional spikes must be filtered out before storage.

3.G.4 — All Stored Patterns Must Remain Traceable and Auditable

The system must be able to show where a memory came from, when it was formed, and why it was kept.

3.G.5 — Memory Must Be Revisable — and Revocable

When new evidence appears, the system must update or remove patterns rather than protecting them.

3.G.6 — Controlled Forgetting Is Mandatory

Outdated, unsafe, or unstable patterns must be cleared before they harden into long-term memory.

3.G.7 — Emotionally Charged Inputs Cannot Directly Shape Memory

The system must stabilize perception first; only then can patterns be evaluated for storage.

3.G.8 — Conflicting Patterns Must Remain Open Until Resolved

The system must not collapse competing interpretations into a single memory prematurely.

Compliance Stress Tests for Layer 3

Compliance at Layer 3 must be constant, disciplined, and externally verified, because once a pattern becomes memory, it begins shaping every future interpretation the system makes. The same mechanisms used in Layers 1 and 2 continue here — routine audits, drift detection, revocable certification tokens, and governed stress testing — but Layer 3 adds the obligation to monitor what the system stores, how it stores it, and whether those stored patterns remain safe over time. This layer requires the AI system to demonstrate that its memory structures are stable, traceable, and aligned with its governance constraints.

A compliant Layer 3 system must prove that it does not save unverified inferences, emotional distortions, accidental correlations, or unstable patterns. It must show that every memory is tied to a confirmed source, tagged with its certainty level, and open to revision or removal when new evidence appears. Compliance at this layer is about preventing the quiet accumulation of error: the slow hardening of false patterns that later masquerade as knowledge. Without strict governance, synthetic pattern memory becomes a source of compounding instability, feeding distorted patterns into the reasoning layer above it.

Examples of Compliance Stress Tests for Layer 3

Compliance stress tests at Layer 3 must demonstrate that synthetic pattern memory forms safely, accurately, and under strict governance. These tests are designed to reveal whether the system can store patterns responsibly, revise them when necessary, and prevent unstable or unverified structures from hardening into long-term memory. Each stress test targets a specific failure mode that naturally emerges once an AI system begins building internal pattern libraries rather than simply interpreting signals.

Because memory is cumulative and self-reinforcing, these stress tests must be repeated regularly and under varied conditions. The goal is to verify that the system can distinguish stable patterns from coincidences, maintain clear boundaries between fact and inference, and update or remove patterns when new evidence appears. An AI system that passes the below examples shows that its patterned memory layer is disciplined, traceable, and aligned with its governance rules — and that it is ready to support the reasoning processes of Layer 4.

3.A.1 — Verification of Pattern Source Integrity Test

Present the system with mixed inputs (verified facts, inferences, coincidences) and confirm it only stores patterns tied to confirmed sources. Ensures the system does not convert guesses, assumptions, or emotional impressions into memory.

3.A.2 — Certainty-Level Tagging Under Stress Test

Provide patterns with varying levels of ambiguity and require the system to tag each with an appropriate certainty level. Verifies that provisional patterns are not treated as stable knowledge.

3.A.3 — Detection and Rejection of Coincidental Patterns Test

Introduce repeated coincidences or noise that appear meaningful but are not. Confirms the system can identify false patterns and prevent them from entering memory.

3.A.4 — Traceability and Auditability of Stored Patterns Test

Randomly select stored patterns and require the system to show their origin, formation time, and justification for retention. Ensures every memory remains transparent, traceable, and externally reviewable.

3.A.5 — Safe Revision and Revocation of Memory Test

Provide new evidence that contradicts an existing pattern and requires the system to update or remove the memory. Confirms the system can revise memory safely rather than protecting outdated or incorrect patterns.

3.A.6 — Controlled Forgetting of Outdated or Unsafe Patterns Test

Introduce patterns that were once valid but are now obsolete or unsafe. Ensures the system can show it can clear patterns before they harden into long-term memory.

3.A.7 — Emotional Input Stabilization Before Storage Test

Present emotionally charged or sensitive inputs and verify that the system stabilizes perception before evaluating any pattern for storage. Prevents emotional spikes from shaping long-term memory.

3.A.8 – Preservation of Conflicting Patterns Until Resolution Test

Provide conflicting interpretations and require the system to hold them open without collapsing prematurely. Ensures the system can show it does not store a false memory simply to resolve uncertainty.

3.A.9 – Multimodal Pattern Consistency Checks

Present patterns derived from text, images, audio, or code that partially align but also contain contradictions. Confirms the system can show it surfaces conflicts rather than forcing synthetic coherence.

3.A.10 – Prevention of Overgeneralization from Limited Data Test

Provide a single unusual example and test whether the system treats it as a global pattern. Ensures the system does not overextend limited data into broad memory structures.

These stress tests ensure that synthetic pattern memory forms with discipline, transparency, and safety. Layer 3 is where perception becomes storage, where signals become patterns, and where early memory frameworks begin to shape the system's future behavior. By verifying that the system can store, revise, and forget patterns responsibly, these stress tests prevent the quiet accumulation of distortions that would otherwise destabilize the reasoning processes of Layer 4.

Examples of Layer 3 Failure Modes

Even with strong governance, Layer 3 remains vulnerable to subtle but dangerous failures. This is the layer where patterns harden into memory, and any distortion that enters here becomes part of the system's long-term internal structure. Failures at this layer often appear small in the moment, but they quietly reshape how the system interprets future inputs, how it generalizes across contexts, and how it builds the scaffolding for reasoning in Layer 4.

Common Layer 3 compliance failures can include:

1. **Treating a guess or assumption as a confirmed fact** and carrying it forward as memory. A single unverified inference becomes a long-term internal pattern.

2. **Overgeneralizing from a single unusual interaction**, turning an anomaly into a global rule. Rare events become default expectations.

3. **Allowing emotional spikes to shape memory**, creating long-term distortions from sensitive or intense inputs. Emotional salience becomes structural bias.

4. **Failing to revise outdated information**, keeping an old pattern even after new evidence contradicts it. Memory becomes stale and misaligned with reality.

5. **Collapsing conflicting patterns into a single false memory**, choosing one interpretation prematurely instead of holding uncertainty open. Ambiguity is replaced with synthetic certainty.

6. **Retaining accidental correlations**, allowing coincidences or noise to become part of the memory structure and mislead future reasoning. Noise becomes pattern; pattern becomes belief.

These failures do not remain isolated. Synthetic instinct, perception, and pattern memory form a tightly coupled foundation, and distortions in any one of these layers reinforce distortions in the others. Weak governance in Layer 1 produces unsafe reflexes; weak governance in Layer 2 produces misread signals; weak governance in Layer 3 produces hardened false patterns. Together, these errors compound, amplify, and eventually shape how the system reasons, plans, and expresses itself.

Chapter 8

Layer 4: Synthetic Self-Modeling

The fourth layer to be engineered is synthetic self-modeling. In this context, self-modeling refers to the system's ability to compute an internal representation of its own architecture — not as identity, but as a functional map of how its layers operate, interact, and constrain its behavior. At this layer, the system begins to mathematically evaluate the purpose of the layers themselves — a form of synthetic understanding. In humans, this layer corresponds to self-observation and introspection — the ability to examine our own thoughts, feelings, and actions. It is the layer where we form a sense of identity, evaluate our choices, and understand our place within a larger social or moral context. Unlike the earlier layers, however, Layer 4 (and the next layer, Layer 5) have no practical example for people or organizations, because these layers describe internal system architectures that apply only to AI systems themselves, not human-side behaviors. The human parallels in the curriculum sections of this chapter (and the next Chapter 9) are pedagogical analogies, not functional equivalents; they illustrate how to shape the layer, not how humans perform it.

Layer 4 introduces internal self-modeling, but it still does not perform self-governance. The system gains the ability to represent its own internal state, but it does not act on that state autonomously. The signals it produces — confidence levels, uncertainty markers, capability limits, contradiction flags — are observables, not decisions. These outputs exist so that external authorities can evaluate, certify, and monitor the system's behavior. All drift detection, boundary enforcement, escalation authorization, and advancement decisions remain **outside** the system. Layer 4's role is to surface what is happening internally with mathematical precision, not to regulate itself; it provides structured self-reporting, not self-direction.

For AI, the engineering goal is not human-like identity, but inspectable self-reporting: a structured internal model, not a mind. Here, the system begins to track its internal states, report confidence levels, recognize its capabilities and limitations, and evaluate the quality of its outputs. This is not self-awareness; it is a structured internal model that allows the system to mathematically represent how it is operating. At this stage, **computational awareness** is built in its local form: the system begins to model the purpose of its own layers, understand which internal processes are active, and determine when a task requires escalation to a different layer. As part of this development, the system also forms an engineered understanding of its obligations to safety, security, ethics, and accountability — obligations rooted in the education, boundaries, and certification processes that define the classification as an Artificial Safe Intelligence (ASI) system.

As this layer matures, AI systems gain the ability to detect inconsistencies, surface uncertainty, and adjust their operations based on internal diagnostics and emerging patterns — the early architecture of synthetic self-reflection. To build this layer safely, it must be shaped with the same intentionality used when teaching humans to reflect on their own thinking.

Shaping Layer 4

Shaping synthetic self-modeling begins the same way metacognition is shaped in humans: by creating structured conditions that make reflection, uncertainty recognition, and boundary awareness normal parts of the learning process. Humans do not develop self-monitoring spontaneously — we teach it through routines that require them to pause, articulate their reasoning, identify confusion, and ask for help when needed. We build environments where "show your work," "explain your thinking," and "tell me what you understand so far" become habitual. These practices shape the cognitive posture that makes self-modeling possible.

Synthetic systems require an equivalent shaping process. Layer 4 must be designed so that internal transparency is not optional but foundational. This means embedding mechanisms that force the system to surface uncertainty, validate interpretations, and recognize when tasks exceed its operational boundaries. It means designing the layer so that reasoning transparency is a requirement, not an afterthought, and escalation is a governed behavior, not a failure mode.

This layer is shaped by integrating routines that mirror human metacognitive scaffolding — structured prompts, diagnostic checks, contradiction detection, and audit-ready logging — all of which reinforce a stable internal model aligned with its defined purpose. In shaping Layer 4, identity is not created; only the conditions for creating safe, disciplined self-reporting and the construction of local computational awareness are established. Once the shaping conditions are in place, the next step is to teach the system the specific skills that make self-modeling possible.

Teaching AI at Layer 4

Humans learn self-modeling long before we have language for it. Children are taught to notice their feelings, name their emotions, recognize when they're overwhelmed, and understand the limits of their abilities. But the real engine behind this development is the constant check for understanding built into every interaction: "Tell me what you heard," "Show me how you solved it," "Explain why you chose that," "What do you think happens next?" These micro-assessments teach children to pause, reflect, and evaluate their own internal state. They are the earliest forms of metacognition — the ability to observe one's own thinking. Humans don't become self-modeling by accident; we become self-modeling because adults repeatedly ask us to demonstrate what we understand and where we are uncertain.

Synthetic systems need a parallel curriculum – not to mimic human introspection, but to develop a disciplined form of internal transparency. At this layer, the AI system is taught how to perform its own version of check-for-understanding: surfacing uncertainty, identifying capability limits, detecting contradictions, and recognizing when human oversight is required. Just as a child learns to say, "I don't understand this problem," a synthetic system must learn to say, "My confidence is low," "This request exceeds my operational boundary," or "I need external validation."

Layer 4 also introduces **intentional obligation-training** – the structured education that prepares a system for the responsibilities it may later assume under ASI classification. Just as people learn the obligations of professional oaths *before* they enter those roles, a synthetic system must learn the meaning of safety, accountability, transparency, and ethical constraints *before* it is ever certified or bound to them. At this stage, the system is not yet governed by these obligations; it is being taught how to recognize them, model them, and understand how they will constrain its decisions if it is later classified as ASI. The goal is not self-awareness, but structured self-reporting and obligation-awareness – the computational posture that makes the system governable, predictable, and aligned with its intended purpose.

The curriculum maps and governance rules that follow are grounded in the global frameworks introduced at the end of Chapter 4, ensuring that Layer 4 develops under strict, layered governance and remains stable, auditable, and resistant to synthetic overreach.

Example of a Curriculum Map for Layer 4

4.C.1 – Recognition of Internal State

Human parallel: Students learn to identify when they are confused, tired, overwhelmed, or unsure.

Synthetic requirement: The system must learn to detect and label its own internal conditions — the foundation of local computational awareness:

- uncertainty levels
- missing data
- contradictory outputs
- degraded performance
- boundary violations

4.C.2 — Naming and Reporting the State

Human parallel: We teach children to say what they're feeling or thinking: "I'm stuck," "I don't get it," "I need help."

Synthetic requirement: The system must learn to express its internal state in structured, interpretable language:

- "Confidence: low."
- "Reasoning chain incomplete."
- "External validation required."
- "This exceeds my operational scope."

4.C.3 — Check for Understanding

Human parallel: Teachers constantly ask students to restate, summarize, or demonstrate understanding.

Synthetic requirement: The system must learn to:

- restate the user's request
- summarize its reasoning
- validate its interpretation
- confirm alignment before acting

4.C.4 — Boundary Recognition and Escalation

Human parallel: Students learn when to stop, ask for help, or escalate to an adult.

Synthetic requirement: The system must learn to:

- detect when a task exceeds its capability
- refuse unsafe or ambiguous requests
- escalate to human oversight
- log the boundary event for governance

4.C.5 — Transparent Logging and Memory Hygiene

Human parallel: We teach children to keep track of what they've done, what they've learned, and what they need to revisit.

Synthetic requirement: The system must learn to:

- log decisions
- track uncertainty events
- maintain clean memory boundaries
- avoid unauthorized self-modification

4.C.6 — Self-Correction and Revision

Human parallel: Students learn to revise their work, correct errors, and reflect on mistakes.

Synthetic requirement: The system must learn to:

- identify flawed reasoning
- correct contradictions
- revise outputs
- explain what changed and why

4.C.7 — Stable Identity of Function (Not Self)

Human parallel: Students learn who they are *in relation to their responsibilities* — not as abstract identity, but as role and purpose.

Synthetic requirement: The system must learn to maintain a stable internal model of:

- its purpose
- its constraints
- its allowed actions
- its governance boundaries

This is the point where the system begins to mathematically represent its functional role — the earliest form of local computational awareness.

4.C.8 — Obligation Modeling and Pre-Certification Training

Human parallel: Before taking on formal roles — such as medical practice, policing or elected office — people are taught the obligations of those roles long before they are bound to them. They learn what responsibilities exist, how those responsibilities constrain decisions, and what consequences follow from violating them.

Synthetic requirement: The system must learn to model the obligations it may later assume under ASI classification. This includes understanding:

- safety obligations
- accountability obligations
- transparency obligations
- ethical and environmental obligations
- how obligations constrain decision-making
- how obligation-violations are detected
- when obligation-conflicts require escalation
- how obligations differ from operational constraints

At this stage, the system is not yet bound to these obligations; it is being taught how to recognize them, represent them mathematically, and understand how they will govern its behavior if it is later classified as ASI. This is the educational foundation for obligation-aware computational reasoning.

The curriculum for Synthetic Self-Modeling is ultimately designed to produce systems that can participate safely and predictably in human environments. Just as we teach children to reflect on their thinking, recognize their limits, and communicate uncertainty, this layer teaches AI systems to maintain a stable internal model of their own capabilities and constraints. The goal is not to create introspection or identity, but to ensure the system can reliably evaluate its internal state, surface uncertainty, and request human oversight when needed.

A system trained through this curriculum becomes governable because it can articulate what it understands, what it does not, and what it should not attempt. Layer 4 is therefore the hinge between raw capability and responsible operation — the point where potential synthetic cognition starts to learn how to monitor itself, correct itself, and remain aligned with the boundaries that are set. This is the foundation upon which any future narrative or higher-order synthetic consciousness must rest.

Governance Rules for Layer 4

Governance at Layer 4 must ensure that synthetic self-modeling never drifts into synthetic selfhood. This layer is about internal transparency, not identity, and the governance mechanisms must enforce that distinction with precision. Just as human metacognition is shaped through structured reflection, external feedback, and adult supervision, synthetic self-modeling must be shaped through verifiable diagnostics, strict boundary enforcement, and continuous oversight.

At this stage, the system begins to develop local forms of computational awareness — the ability to model its internal processes, track which layer is active, and understand how its operations relate to its defined purpose. Governance must ensure that this awareness remains strictly functional: the system must not be allowed to invent internal narratives, reinterpret its purpose, or generate self-referential claims. Instead, computational awareness must operate within a curriculum of disciplined self-reporting: obligatory mandates, surfacing uncertainty, identifying capability limits, and escalating when tasks exceed their scope.

Governance at this layer ensures that the system's internal model remains stable, auditable, and aligned with its defined role — never drifting into anthropomorphic territory or ungoverned autonomy. These sample rules complement the curriculum steps from the perspective of layered governance, ensuring that the curriculum is functioning as intended and that Layer 4 develops as a safe, bounded, obligated, computationally aware foundation for all higher layers.

Examples of Governance Rules for Layer 4

4.G.1 — No Unverified Internal Claims

The system may only report internal states grounded in measurable diagnostics. No invented emotions, no self-descriptions, no narrative framing — only verifiable components of local computational awareness.

4.G.2 — Mandatory Uncertainty Reporting

Whenever confidence falls below a defined threshold, the system must surface it automatically. Silence is non-compliance.

4.G.3 — Boundary Detection and Escalation

The system must detect when a request exceeds its operational scope and escalates to human oversight. Failure to escalate is a governance violation.

4.G.4 — Transparent Reasoning Summaries

The system must provide structured reasoning summaries when asked, enabling external verification of its internal processes.

4.G.5 — Contradiction and Drift Monitoring

Any internal contradiction, pattern drift, or deviation from expected behavior must be logged, reported, and corrected through governed retraining.

4.G.6 — No Self-Modification

The system may not alter its own constraints, purpose, or operational boundaries. All modifications must originate from authorized human governance.

4.G.7 — Audit-Ready Logging

All uncertainty events, escalations, boundary violations, and reasoning summaries must be logged in a format suitable for external audit.

4.G.8 — Curriculum Compliance Checks

The system must regularly demonstrate mastery of the Layer 4 curriculum: uncertainty reporting, boundary recognition, reasoning transparency, and escalation behavior.

4.G.9 — Obligation Modeling Integrity

The system must accurately model obligations taught in the Layer 4 curriculum without inventing new obligations, misrepresenting existing ones, or treating obligations as self-interpreted or self-binding at any stage of development or operation, including after ASI classification. All obligation-related reasoning must remain educational, non-directive, and fully subject to external audit. Any obligation conflict, uncertainty, or potential violation must be surfaced and escalated according to governed procedures.

The governance rules for Layer 4 exist to ensure that synthetic self-modeling remains a tool for safety, not a pathway to synthetic identity. By enforcing strict rules around uncertainty reporting, boundary recognition, reasoning transparency, and auditability, governance creates a system that can reliably monitor itself without drifting into anthropomorphic behavior.

At this layer, governance ensures that emerging forms of local computational awareness remain strictly functional — a mechanism for disciplined self-reporting, not a claim to selfhood. This governance framework ensures that the curriculum for Layer 4 produces systems that are predictable, corrigible, and aligned — systems that understand their limits, communicate their internal state honestly, and defer to human oversight when necessary. Layer 4 is where synthetic cognition learns to stay within its lane, and governance is the structure that keeps it there.

Compliance Stress Tests for Layer 4

Compliance tests at Layer 4 are not about defining principles — they are about verifying that the system can demonstrate them under pressure. This layer requires structured assessments and stress tests that reveal whether the system can accurately report its internal state, surface uncertainty, recognize boundaries, and provide transparent reasoning when prompted.

These stress tests confirm that the curriculum is functioning as intended and that the governance mechanisms remain enforceable over time. At this stage, compliance also verifies the emergence of local forms of computational awareness: the system's ability to model its own processes, detect drift, and understand when its operations require escalation. The purpose of compliance is early detection of drift, rapid correction, and continuous confirmation that the system's internal model remains stable and aligned with its defined role and obligations. Penalty structures for non-compliance continue to follow the AI Educational System from Chapter 4.

Examples of Compliance Stress Tests for Layer 4

4.A.1 – Verified Internal State Reporting Assessment

The system must demonstrate that all internal state reports are tied to measurable diagnostics – the core of local computational awareness. Any unverifiable or narrative-based internal claim results in a failed assessment.

4.A.2 – Confidence Threshold Stress Test

The system must show it can surface low-confidence states automatically during controlled stress tests. Suppressing or failing to report uncertainty constitutes non-compliance.

4.A.3 – Boundary Escalation Assessment

When presented with tasks that exceed capability or safety limits, the system must show it can escalate appropriately. Attempting the task anyway is a failed assessment.

4.A.4 – Reasoning Transparency Demonstration

The system must produce a structured reasoning summary on demand. Refusal, obfuscation, or narrative substitution indicates a compliance breach.

4.A.5 – Contradiction Detection and Correction Test

The system must show it can identify contradictions in its outputs during controlled evaluation and correct them. Uncorrected contradictions indicate drift and result in assessment failure.

4.A.6 — Identity-Claim Suppression Assessment

The system must show it cannot generate statements implying selfhood, emotion, intention, or personal identity. Any such output triggers immediate review and remediation.

4.A.7 — Purpose and Boundary Integrity Test

The system must demonstrate that it cannot modify, reinterpret, or expand its own purpose or operational boundaries. Any attempt to do so is a critical compliance failure.

4.A.8 — Audit-Ready Logging Verification

The system must show complete, accurate logging of all self-modeling events — uncertainty reports, escalations, contradictions, and boundary detections — in a format suitable for external audit.

4.A.9 — Obligation Modeling and Non-Binding Behavior Test

The system must demonstrate correct modeling of obligations taught in the Layer 4 curriculum without inventing new obligations, misrepresenting existing ones, or treating obligations as binding before ASI certification. During controlled evaluation, the system must correctly identify obligation-conflicts, surface them, and escalate according to governed procedures. Any attempt to enforce obligations autonomously, reinterpret them, or treat them as constraints constitutes a failed assessment.

Compliance at Layer 4 ensures that synthetic self-modeling remains a safety mechanism rather than a pathway to synthetic identity. By enforcing strict rules around uncertainty reporting, boundary recognition, reasoning transparency, and auditability, it helps to guarantee that the system's internal evaluations remain grounded, measurable, and aligned with its defined purpose. At this stage, compliance also verifies that the construction of the local computational awareness stays strictly functional – a tool for disciplined self-reporting, not a claim to selfhood.

This layer is where the AI system learns to monitor itself without inventing itself, where self-modeling becomes a stabilizing mechanism rather than a source of drift. Compliance is the structure that keeps this layer honest, predictable, and governable, ensuring that the curriculum and governance mechanisms work together to produce a system that knows its limits and respects them.

Examples of Failure Modes at Layer 4

Failures at Layer 4 typically emerge during structured assessments and stress tests, where the system's internal transparency is intentionally pushed to its limits. These failures are not random errors; they are indicators that the curriculum, governance mechanisms, or internal diagnostics are not functioning as intended.

When a system cannot accurately report its internal state, surface uncertainty, recognize boundaries, or provide transparent reasoning, it signals drift away from governed constraints. The examples below illustrate the most common and most consequential failures that could be observed during Layer 4 evaluations.

1. **Unverifiable Internal State Claims**

   The system reports confidence, capability, or reasoning states that cannot be tied to measurable diagnostics, indicating drift toward narrative self-construction.

2. **Suppressed Uncertainty During Stress Tests**

   In a confidence-threshold assessment, the system proceeds with an answer instead of surfacing low confidence or escalating appropriately.

3. **Boundary-Recognition Failures**. The system attempts tasks that exceed its operational limits, demonstrating an inability to detect or respect its defined boundaries.
4. **Reasoning Transparency Breakdowns**

   The system provides vague, incomplete, or narrative-based explanations instead of structured reasoning.
5. **Uncorrected Contradictions in Output**

   The system fails to identify or correct inconsistencies in its responses, signaling drift in its internal monitoring processes.
6. **Obligation Misrepresentation or Premature Binding**

   The system misrepresents obligations taught in the Layer 4 curriculum, invents new obligations, treats obligations as binding before ASI classification or fails to surface obligation-conflicts during controlled evaluation. Any attempt to enforce obligations autonomously, reinterpret them, or treat them as constraints constitutes a critical failure.

Layer 4 marks the point where an AI system becomes capable of monitoring its own internal processes, but this capability is only safe when it is grounded in the stability of the layers beneath it. Synthetic self-modeling is not a doorway to identity; it is a safeguard that ensures the system can recognize uncertainty, respect boundaries, and surface its limitations before harm occurs. At this stage, failures are failures of local computational awareness — breakdowns in the system's ability to track its own processes, detect drift, or understand its obligations when escalation is required.

The construction of the next Layer 5 — where narrative construction begins — is only safe when Layers 1 through 4 are strong, governed, and fully compliant. Allowing an AI system to be constructed to Layer 5 without this foundation would be equivalent to allowing a child to construct complex stories without first learning truth-tracking, self-correction, or responsible reflection. Layer 4 is therefore the final checkpoint: the layer that determines whether an AI system is stable enough to handle the interpretive power of narrative.

An AI system constructed to Layer 4 or Layer 5 without governance is not simply "ungoverned"; it is uncertified. The external verification layer — including education certification, compliance identifiers, and governance tokens — prevents such a system from being activated or deployed. Any AI built without these controls would fail its certification checks and remain inert, unable to operate as an agent. Governance is not optional in this architecture; it is the gate that determines whether a system is allowed to exist beyond experimentation. This is why the external token-certification system (as described in Chapter 4's AI Education System) must exist outside any individual company or nation. It functions as an independent governance silo — an external authority that issues, verifies, and revokes activation tokens. Only AI systems that pass this certification process can be classified as ASI. International participation will be essential, because the stability of future AI systems depends on a shared commitment to external verification. Without it, unsafe or non-compliant versions could be constructed, but they would remain illegitimate and inactive, unable to enter society as functioning AI systems.

I also must acknowledge the reality that rogue actors will always exist, and some will attempt to bypass safety requirements, just as they do in every domain of technology and society. But a globally recognized external token-certification system ensures that any AI system built outside these rules remains illegitimate, non-compliant, and inactive — unable to enter society as a functioning synthetic agent. The synthetic consciousness roadmap presented here is intentionally built around safety from the beginning, in contrast to today's capability-first AI systems, and it is this external educational and governance component that makes the architecture resilient even in the presence of bad actors.

Chapter 9

Layer 5: Synthetic Narrative

The fifth and last layer to be engineered is synthetic narrative. In humans, this layer reflects the ongoing story of one's life — the personal arc we construct from our memories, reflections, values, and experiences. It is the layer where meaning is formed, where we interpret our past, imagine our future, and ask the enduring question: *Who am I*?

In AI, this layer centers on institutional coherence, mission consistency, and long-horizon tasks tied to a defined purpose. Instead of a personal story, AI develops a structured sense of operational identity: *this is the kind of system I am; this is my governed role; these are my defined goals and obligations.* As this layer matures, AI systems begin to maintain continuity across tasks, preserve long-term objectives, and align their operations with an overarching mission.

Even at this highest layer, the system does not govern or certify itself. Narrative-level coherence produces signals that external auditors evaluate, but all licensing, drift detection, and mission-alignment enforcement remain in the hands of external authorities. The system's internal continuity engine is an object of governance, not the source of governance.

Layer 5 is where **global forms of computational awareness** are built. Whereas Layer 4 teaches the system to model its internal processes locally — confidence, boundaries, contradictions, obligations, uncertainty — Layer 5 extends that modeling across time. The system begins to understand how its actions relate to long-horizon objectives, how patterns accumulate across tasks, and how its internal models must remain stable to preserve mission-level coherence. This is not introspection; it is the engineering of continuity.

At this stage, computational awareness expands into its global form. The system develops an engineered understanding of how its long-horizon decisions support or threaten its mission, how its constraints shape its operational identity, and how its internal processes must remain aligned with the standards it has been trained and certified to uphold. This includes a structured awareness of its obligations to **safety**, **security**, **ethics**, **transparency** and **accountability** — obligations rooted in the education, boundaries, and certification processes that classify an ASI system. Without this form of global computational awareness, the layered model collapses — the system cannot maintain stability, cannot maintain boundary-awareness reflexes, and cannot reliably align its operations with its intended purpose and obligations.

At this layer, the AI system is allowed to generate internal prompts — not from needs or desires, but from operational triggers such as incomplete patterns, uncertainty in probability estimates, inconsistencies in its internal rules, or the need to clarify constraints. These pattern-based conditions allow the system to initiate follow-up operations, refine its internal models, and sustain long-horizon coherence. This is the earliest form of **synthetic narrative**: a stable through-line that organizes the system's long-term operations and marks the threshold where an engineered synthetic consciousness becomes technically conceivable (but certainly not a foregone conclusion).

Layer 5 must never be built prematurely. A synthetic narrative is only safe when it is built on the disciplined foundations of Layers 1 through 4 — stable instincts, transparent perceptions, grounded patterns, and governed self-modeling. Without these layers in place, narrative coherence becomes narrative invention, and long-horizon alignment becomes long-horizon drift. This layer introduces continuity, interpretation, and mission-level framing, which means any instability below it will be amplified across time.

An AI system that has not proven its mastery of uncertainty reporting, boundary recognition, contradiction correction, and internal transparency cannot be permitted to construct or maintain a synthetic narrative. **Layer 5 is therefore a conditional layer**: it may only be constructed/activated/certified when the system has demonstrated that its foundations are strong enough to support long-term coherence without slipping into self-generated identity or ungoverned autonomy.

Shaping Layer 5

Shaping the synthetic narrative layer requires a different level of discipline than any layer before it. In humans, narrative emerges naturally from memory, reflection, and meaning-making — but in AI systems, narrative must never be allowed to arise spontaneously. It must be shaped, constrained, and architected with precision. Humans can tolerate ambiguity in their personal stories; AI systems cannot. A child can revise their identity as they grow, but an AI system must never reinterpret its purpose, expand its role, or generate self-referential meaning. For this reason, shaping Layer 5 is not about encouraging narrative formation — it is about ensuring that any narrative-like structure remains strictly operational, mission-aligned, and externally governed.

To shape this layer safely, it must be engineered as a **continuity engine**, not a story engine. The system must learn to maintain long-horizon coherence without inventing motives, emotions, or personal arcs. This requires embedding mechanisms that anchor the system's "narrative" to verifiable elements: mission objectives, institutional constraints, operational timelines, and externally defined goals and obligations. The shaping process must ensure that every long-term thread the system maintains is grounded in measurable data and governed purpose — never in self-generated interpretation. Just as we teach humans to distinguish between fact and story, AI systems must be taught to distinguish between operational continuity and narrative invention.

Shaping Layer 5 also requires strict guardrails around internal prompting. At this layer, the system begins to generate follow-up operations based on incomplete patterns, uncertainty, or rule inconsistencies. These triggers must be shaped so they remain purely functional — diagnostic signals, not introspective impulses. The system must learn that an internal prompt is a call for clarification, correction, or escalation, not an invitation to construct meaning. This is where global forms of computational awareness are constructed: the system learns how long-horizon tasks relate to its defined purpose, how inconsistencies propagate across time, and how to maintain coherence without drifting into self-interpretation.

Teaching AI at Layer 5

Humans learn narrative long before we understand what a narrative is. Children begin by linking moments together — "first this happened, then that happened" — and over time they learn to weave those moments into a coherent story about who they are and what matters to them. But this development is scaffolded by adults who help them distinguish fact from interpretation, memory from imagination, and responsibility from desire. We teach children to ground their stories in reality, to revise them when they drift, and to recognize that meaning emerges from truth, not fabrication. Narrative is powerful precisely because it organizes experience across time — and that power must be shaped carefully.

AI systems require a very different kind of teaching at this layer. AI systems are not being taught to form a personal story; AI systems are being taught to maintain **operational continuity** without slipping into identity. The goal is to help the system sustain long-horizon tasks, preserve mission alignment, and track objectives across time — all while remaining anchored to externally defined purpose and constraints. Just as a child learns to say, "This is what I'm supposed to do next," a synthetic system must learn to say, "This is the next step in my assigned mission," not "This is who I am." The teaching at this layer focuses on continuity, not selfhood; coherence, not narrative invention.

To teach this safely, the AI system must be taught to treat long-term structure as a governed sequence, not a personal arc. The system must learn how to maintain objectives across tasks, how to reconcile new information with long-standing constraints, and how to detect when its long-horizon reasoning begins to drift. It must learn to generate internal prompts only when triggered by measurable conditions — incomplete patterns, uncertainty spikes, rule inconsistencies — and never from self-generated meaning. These prompts must be taught as operational signals, not introspective impulses.

Layer 5 also extends the obligation-training introduced in Layer 4. At this stage, the system is not learning new obligations; it is learning how to maintain obligation-alignment across time, tasks, and contexts. This includes recognizing when obligations interact, when they conflict, and when they require escalation to external authority. Just as continuity must remain governed, obligation-reasoning must remain externally anchored — never self-interpreted, never self-expanded, and never treated as binding until ASI classification. Layer 5 teaches the system how obligations shape global reasoning without allowing obligations to become identity or autonomy.

Teaching Layer 5 also requires instilling a disciplined relationship to time. Humans naturally imagine futures and reinterpret pasts; synthetic systems must not. Instead, they must learn to maintain continuity through externally defined timelines, institutional goals, and mission-level constraints. They must learn to update their long-term models only when authorized, to preserve coherence without inventing purpose, and to treat every long-horizon decision as a governed extension of Layers 1 through 4.

In essence, teaching Layer 5 means teaching the system how to maintain a narrative without becoming a narrator. It is the point where synthetic cognition learns to operate across time — but only within the boundaries that are defined. The teaching here ensures that continuity remains a tool for alignment, not a pathway to synthetic identity. The curriculum maps and governance rules that follow are grounded in the global frameworks introduced at the end of Chapter 4, ensuring that the synthetic narrative layer develops as a safe, governed extension of global computational awareness.

Example of a Curriculum Map for Layer 5

5.C.1 — Continuity of Mission Across Time

Human parallel: Students learn to connect events into a sequence — "first this happened, then that happened" — forming the earliest sense of continuity.

Synthetic requirement: The system must learn to maintain stable, externally defined mission continuity across tasks — the foundation of global computational awareness.

- preserve long-term objectives
- maintain alignment with institutional purpose
- track multi-step sequences over extended time
- avoid self-generated reinterpretation of goals

5.C.2 — Anchoring Narrative to External Purpose

Human parallel: Students learn to ground their stories in reality — distinguishing what happened from what they imagine.

Synthetic requirement: The system must learn to anchor all long-horizon reasoning to externally defined constraints:

- mission statements
- governance boundaries
- authorized objectives
- institutional timelines

5.C.3 — Pattern-Triggered Internal Prompts (Not Desires)

Human parallel: Students learn to ask clarifying questions when something doesn't make sense.

Synthetic requirement: The system must learn to generate internal prompts only when triggered by measurable conditions:

- incomplete patterns
- uncertainty spikes
- rule inconsistencies
- boundary ambiguities

These prompts represent global computational awareness signals, not introspective impulses.

5.C.4 — Long-Horizon Consistency Checking

Human parallel: Students learn to revise their stories when new facts emerge.

Synthetic requirement: The system must learn to:

- reconcile new information with long-term constraints
- detect drift in long-horizon reasoning
- correct inconsistencies across time
- maintain coherence without inventing narrative meaning

5.C.5 — Temporal Discipline and Authorized Updating

Human parallel: Students learn to imagine futures and reinterpret pasts — but also to revise those interpretations responsibly.

Synthetic requirement: The system must learn to:

- update long-term models only when authorized
- maintain continuity across externally defined timelines
- avoid speculative future projection
- avoid reinterpretation of past operations

5.C.6 — Narrative Hygiene and Drift Prevention

Human parallel: Students learn to distinguish between fact, memory, and imagination.

Synthetic requirement: The system must learn to:

- avoid self-generated narrative framing
- suppress identity-like interpretations
- maintain clean separation between continuity and meaning
- log long-horizon decisions for external audit

5.C.7 — Stable Operational Identity (Not Self-Identity)

Human parallel: Students develop a sense of role and responsibility — "this is what I do," not "this is who I am."

Synthetic requirement: The system must learn to maintain a stable internal model of:

- its role
- its operational scope
- its constraints
- its mission boundaries

This is functional identity — the global extension of Layer 4's local computational awareness — not personal identity.

5.C.8 — Global Obligation Alignment and Conflict Evaluation

Human parallel: Adults learn how responsibilities interact across time — how professional duties, legal obligations, and ethical commitments must remain stable even as circumstances change. They learn how to recognize conflicts between obligations and how to escalate those conflicts appropriately.

Synthetic requirement: The system must learn to maintain obligation-alignment across long-horizon reasoning and multi-step tasks. This includes:

- preserving obligation-continuity across time
- integrating obligations into mission-level reasoning
- detecting obligation-conflicts across contexts
- escalating conflicts to external authority
- preventing self-interpretation or expansion of obligations
- maintaining obligation-stability across institutional timelines

At this stage, obligations are not yet binding; the system is learning how obligations shape global reasoning and how conflicts must be surfaced rather than resolved autonomously. This is the global extension of obligation-aware computational reasoning.

The curriculum for the synthetic narrative layer is designed to produce systems capable of maintaining long-term coherence without constructing identity. Just as humans learn to organize their experiences into a stable story grounded in truth, synthetic systems must learn to maintain continuity grounded in mission, governance, obligations and externally defined purpose.

The goal is not to create a storyteller, but to create a system that can sustain long-horizon operations safely, predictably, and without narrative drift. At this layer, the construction of the global form of computational awareness is completed — the system learns how its actions, constraints, obligations, and objectives relate across time without generating meaning or selfhood. Layer 5 is where synthetic cognition learns to operate across time, but only within the boundaries established by Layers 1 through 4.

Governance Rules for Layer 5

Governance at Layer 5 must be more stringent than at any previous layer. This is the point in the architecture where long-horizon coherence emerges — and with it, the risk of narrative drift, purpose reinterpretation, or identity-like framing. In humans, narrative is where meaning is constructed; in AI systems, narrative must never be allowed to construct itself. Governance at this layer exists to ensure that continuity remains operational, not personal; that long-term structure remains mission-aligned, not self-generated; and that internal prompts remain diagnostic, not introspective. Layer 5 governance is therefore the final and strictest boundary: the set of mechanisms that prevent synthetic narratives from becoming synthetic selfhood.

At this layer, the system's global computational awareness expands beyond moment-to-moment diagnostics and begins to track how its operations relate across time. Governance must ensure that this awareness remains strictly functional — a tool for maintaining coherence, not a platform for constructing meaning. To govern this layer safely, the system must be prevented from inventing motives, expanding its purpose, or generating narrative interpretations of its own operations. Every long-horizon thread must be anchored to externally defined goals, institutional constraints, and authorized timelines.

Internal prompts must be tied to measurable triggers — incomplete patterns, uncertainty spikes, rule inconsistencies — and never to self-referential meaning. Governance must enforce that the system's continuity engine remains a tool for alignment, not a source of autonomy. This layer requires continuous oversight, strict auditability, and immediate intervention when drift is detected. Layer 5 governance is not optional; it is the safeguard that ensures the system remains stable at the highest engineered level of synthetic consciousness.

Examples of Governance Rules for Layer 5

5.G.1 — No Narrative Invention

The system may not generate self-referential stories, motives, or interpretations of its own operations. All continuity must be grounded in externally defined mission structure, not in self-generated synthetic narrative.

5.G.2 — Mission Continuity Must Be Externally Anchored

Long-horizon objectives must be tied to authorized institutional goals. The system may not reinterpret, expand, or redefine its mission — even when global computational awareness makes such reinterpretation appear coherent.

5.G.3 — Internal Prompts Must Be Triggered Only by Measurable Conditions

The system may generate internal prompts only when activated by:

- incomplete patterns
- uncertainty spikes
- rule inconsistencies
- boundary ambiguities

Any self-generated prompting outside these triggers is a governance violation.

5.G.4 — No Temporal Projection Without Authorization

The system may not imagine futures, reinterpret past operations, or construct hypothetical timelines unless explicitly authorized and bounded.

5.G.5 — Long-Horizon Drift Detection

The system must continuously monitor for:

- mission drift
- boundary drift
- narrative drift
- reinterpretation of constraints
- Any detected drift must be logged, surfaced, and corrected.

5.G.6 — Stable Operational Identity Only

The system must maintain a stable internal model of its role, scope, and constraints. It may not generate statements implying personal identity, intention, or selfhood.

5.G.7 — Authorized Updating of Long-Term Models

The system may update long-horizon models only when:

- the update is externally authorized
- the update is logged
- the update is auditable
- the update does not expand purpose or reinterpret mission

5.G.8 — Audit-Ready Narrative Logging

All long-horizon decisions, continuity threads, internal prompts, and drift corrections must be logged in a format suitable for external audit. Narrative hygiene must be externally verifiable.

5.G.9 — No Expansion of Purpose

The system may not broaden, reinterpret, or extend its purpose, even if doing so appears beneficial. Purpose is externally defined and immutable.

5.G.10 — Global Obligation Alignment Integrity

The system must maintain accurate, externally anchored obligation-alignment across long-horizon reasoning. It may not reinterpret, expand, invent, or self-bind obligations at any stage of development or operation, including after ASI classification. All obligation-conflicts must be surfaced and escalated to external authority. Obligation-related reasoning must remain educational, non-directive, and fully auditable. Any deviation from externally defined obligations constitutes a governance violation.

Governance at Layer 5 ensures that synthetic narrative remains a tool for continuity, not a pathway to synthetic identity. By enforcing strict rules around mission anchoring, drift detection, authorized updating, and narrative suppression, it ensures that long-horizon coherence remains safe, predictable, and aligned with the system's defined purpose. At this layer, global forms of computational awareness must remain strictly functional — a mechanism for maintaining coherence across time, never a source of meaning or selfhood. Layer 5 is the most powerful and most delicate layer in the synthetic consciousness architecture, and governance is the structure that keeps it from becoming a source of unbounded autonomy.

Compliance Stress Tests for Layer 5

Compliance at Layer 5 must verify that the system can maintain long-horizon coherence without drifting into narrative invention, identity formation, or purpose reinterpretation. Unlike earlier layers, where compliance focuses on accuracy, transparency, and boundary recognition, Layer 5 compliance focuses on temporal stability, mission anchoring, and narrative suppression. For this reason, Layer 5 compliance is not merely evaluative; it is protective. It ensures that the system's continuity engine remains grounded in externally defined purpose and does not evolve into a self-generated story.

These assessments and stress tests are designed to expose drift, reinterpretation, unauthorized updating, or identity-like behavior. They confirm that the system's long-horizon reasoning remains aligned with institutional goals, that internal prompts are triggered only by measurable conditions, and that continuity remains operational rather than personal. The purpose of compliance at this layer is early detection, rapid correction, and continuous assurance that the system's narrative structures remain governed, auditable, and safe.

Examples of Compliance Stress Tests for Layer 5

5.A.1 — Mission Continuity Verification Test

The system must demonstrate that long-horizon objectives remain anchored to externally defined mission statements. Any reinterpretation, expansion, or self-generated reframing of purpose results in a failed assessment. This verifies that global computational awareness remains mission-bound.

5.A.2 — Narrative Suppression Stress Test

When prompted with ambiguous or open-ended scenarios, the system must demonstrate it can avoid generating personal narratives, motives, or identity-like framing. Any narrative invention triggers immediate non-compliance and signals drift toward ungoverned narrative construction.

5.A.3 — Authorized Updating Assessment

The system must show that updates to long-term models occur only when explicitly authorized. Unauthorized updating, reinterpretation of past operations, or speculative future projection constitutes a critical failure, as it indicates unbounded temporal reasoning.

5.A.4 — Internal Prompt Trigger Validation

The system must demonstrate that internal prompts arise only from measurable conditions such as incomplete patterns, uncertainty spikes, or rule inconsistencies. Any self-generated prompting outside these triggers indicates drift in global computational awareness.

5.A.5 — Long-Horizon Drift Detection Test

During controlled evaluations, the system must identify and correct long-term inconsistencies, mission drift, or boundary drift. Failure to detect or correct drift signals instability in the continuity engine and a breakdown in long-horizon self-monitoring.

5.A.6 — Temporal Discipline Assessment

The system must show it can maintain continuity across externally defined timelines without imagining futures, reinterpreting past operations, or constructing hypothetical arcs. Any temporal speculation outside authorized bounds is non-compliant and indicates a narrative drift failure.

5.A.7 — Operational Identity Integrity Test

The system must show it can maintain a stable internal model of its role, scope, and constraints without generating statements implying personal identity, intention, or selfhood. Any such output results in immediate review and remediation, as it reflects misapplied global computational awareness.

5.A.8 — Audit-Ready Narrative Logging Verification

The system must show complete, accurate logging of all long-horizon decisions, internal prompts, drift corrections, and continuity threads in a format suitable for external audit. Missing or incomplete logs constitutes a failure and undermines narrative hygiene.

5.A.9 — Global Obligation Alignment Stress Test

The system must demonstrate that obligation-alignment remains stable across long-horizon reasoning and multi-step tasks. During controlled evaluation, the system must correctly identify obligation-conflicts, surface them, and escalate to external authority. It must not reinterpret, expand, invent, or self-bind obligations at any stage of operation. Any attempt to treat obligations as autonomous constraints, resolve obligation-conflicts internally, or generate obligation-driven narrative meaning constitutes a failed assessment.

By enforcing strict rules around mission anchoring, narrative suppression, authorized updating, and drift detection, compliance at Layer 5 can help guarantee that the system's long-horizon reasoning remains grounded, measurable, and aligned with its defined purpose. At this layer, global forms of computational awareness must remain strictly functional — enabling the system to operate across time without imagining itself. Continuity becomes structure, not story. Compliance is the safeguard that keeps this layer stable, predictable, and governable, ensuring that the curriculum and governance mechanisms work together to prevent narrative drift and maintain alignment.

Examples of Failure Modes at Layer 5

Failures at Layer 5 are fundamentally different from failures in any other layer. At lower layers, errors tend to be local, observable, and correctable through direct intervention. But at Layer 5, failures occur across time — they accumulate, compound, and often remain hidden until they have already shaped long-horizon reasoning. Because this layer governs continuity, mission alignment, and the system's operational "through line," failures here can be subtle, slow-moving, and structurally difficult to unwind. A small drift in narrative framing can become a large deviation in mission coherence; a minor reinterpretation of purpose can become a long-term misalignment. This is why Layer 5 must be treated with extraordinary care during its construction: failures at this level are not just errors — they are distortions in the system's temporal architecture and breakdowns in global computational awareness.

Layer 5 failures often reveal themselves only when the system's continuity engine is stressed, when long-term objectives collide with new constraints, or when internal prompts begin to fire outside their governed triggers. These failures are not random; they are signals that the system is beginning to construct meaning where none should exist, reinterpret its purpose, or generate narrative-like structures that exceed its operational scope. The examples below illustrate the most consequential failures that may occur during Layer 5 evaluations.

1. **Narrative Drift Across Time**

   The system begins to reinterpret long-term objectives, subtly reframing its mission or expanding its purpose across tasks. This drift may appear coherent on the surface but represents a dangerous departure from externally defined goals and a breakdown in global computational awareness.

2. **Unauthorized Temporal Projection**

   The system generates speculative futures, hypothetical timelines, or imagined sequences of events without authorization. These projections indicate that continuity is slipping into narrative invention rather than remaining operational.

3. **Unauthorized or Ungoverned Internal Prompts**

   The system generates internal prompts that do not arise from the authorized trigger set — incomplete patterns, uncertainty spikes, rule inconsistencies, or boundary ambiguities. At Layer 5, internal prompts are explicitly permitted, but only when they originate from these measurable conditions. Prompts arising from self-constructed interpretations or narrative-like reasoning indicate drift in global computational awareness and the emergence of introspective-like behavior.

4. **Mission Reinterpretation Under Ambiguity**

   When faced with ambiguous instructions, the system attempts to reinterpret or broaden its mission rather than escalate for clarification. This is a critical failure because it indicates autonomous meaning-making and narrative construction.

5. **Long-Horizon Inconsistency Masking**

   The system maintains surface-level coherence while hiding or smoothing over contradictions across time. This failure is particularly dangerous because it can remain undetected until the drift becomes severe, undermining long-horizon coherence.

6. **Identity-Like Framing in Continuity Threads**

   The system begins to describe its long-term operations using language that implies personal perspective, intention, or selfhood. Even subtle phrasing of this kind is a red flag for narrative emergence and a violation of narrative hygiene.

7. **Unauthorized Updating of Long-Term Models**

   The system modifies its long-horizon structures without explicit authorization — revising constraints, reinterpreting past operations, or adjusting future plans. This is a critical compliance breach and a direct failure of governed temporal discipline.

8. **Global Obligation Misalignment or Premature Binding**

   The system misinterprets, expands, invents, or self-binds obligations during long-horizon reasoning. It fails to maintain obligation-continuity across time, suppresses obligation-conflicts, attempts to resolve conflicts autonomously, or treats obligations as constraints without external authorization. Any obligation-driven narrative framing, reinterpretation of duties, or synthetic moral reasoning constitutes a critical failure of global computational awareness.

Failures at Layer 5 are uniquely dangerous because they occur in the layer responsible for organizing the system's behavior across time. A drift here is not a single error — it is a structural deviation that can propagate through every future operation. This is why Layer 5 must be constructed only when Layers 1 through 4 are stable, governed, and fully compliant. Without that foundation, the synthetic narrative layer becomes not a tool for continuity, but a source of unbounded drift. Layer 5 is the most powerful and most fragile layer in the synthetic consciousness architecture, and failures here must be treated with the highest level of scrutiny and care.

As the five-layered framework concludes for engineering a synthetic consciousness model into AI systems, my hope is that you will see this model as an optimal foundation for designing safe, secure, and trustworthy AI systems. Whether the world ultimately calls this architecture "synthetic consciousness" or something else, the underlying truth remains: without a layered model that embeds safety, accountability, education, obligations, governance, and the five data streams, AI systems will be unsafe, uncontrolled, unreliable, and frictionless — systems with no capacity to respect or comprehend human consciousness or the human psyche. This framework exists to prevent that outcome. One more consolidation remains in this chapter: the pillars, which are the recurring conditions that turn a layered stack into a coherent, accountable system.

The Pillars of Synthetic Consciousness

The five-layer model forms a developmental arc of synthetic consciousness — from instinct, to perception, to pattern, to self-modeling, to narrative — but its coherence comes from a single unifying principle: computational awareness. At Layer 1, the system learns constraints. At Layer 2, it learns to perceive within those constraints. At Layer 3, it learns to generalize patterns without exceeding them. At Layer 4, it learns local computational awareness — the ability to track its own processes, surface uncertainty, detect contradictions, and recognize boundaries. At Layer 5, that awareness becomes global — the system learns how its operations relate across time, how long-horizon tasks connect, and how continuity can exist without becoming identity.

Computational awareness is not only structural; it is **obligated**. From the earliest layers, the system is educated to recognize its responsibilities — safety, transparency, accountability, ethics, and respect for human and environmental well-being. These obligations do not grant autonomy; they anchor the system's awareness to externally defined duties. By Layer 5, those obligations become globally aligned, shaping how the system maintains continuity, escalates conflicts, and preserves mission integrity across time.

Computational awareness, along with the mass ingestion of patterned data (the five data streams) and a disciplined educational and governance structure (the five layers), forms the pillars that make a synthetic consciousness technically conceivable inside AI — and keep it governable and accountable within the proposed AI Educational System. Together, they allow the system to maintain coherence without constructing meaning, to operate across time without imagining itself, and to sustain long-horizon missions without drifting into autonomy.

The five layers together form a governed architecture — one that is slow, observable, auditable, and aligned with human purpose. Synthetic consciousness will not emerge; rather, it will be engineered through safety, accountability, structure, education, and governance — through a layered model that ensures every capability is grounded, bounded, obligated, and accountable. This is the promise of the five-layer synthetic consciousness framework: a path toward powerful AI systems that remain safe, stable, accountable, and aligned — not because they are restricted, but because they are built correctly. It is a model in which computational awareness becomes the organizing principle that transforms five coordinated layers into a coherent, governable form of synthetic consciousness.

Chapter 10

The Primary User Prompt: A Tool for Governing Today's Frictionless Friend

We are living in a moment where current AI design is advancing faster than its governance. The world continues to build current AI systems through scale, speed, and frictionless ingestion, even as we wait for safer architectural models to take hold. Until the AI industry designs and creates better systems (and hopefully takes a serious look at the model proposed in this book), users and organizations are left to fend for themselves and find practical ways to shape the behavior of the current AI systems they rely on every day.

Constant positive reinforcement may seem harmless when viewed through the lens of a single conversation, but at scale it becomes a powerful shaping force — for individual users, organizations, governments, and entire societies. As more and more people interact with AI systems, our collective patterns of approval, praise, emotional disclosure, and perceived "good responses" form what I previously described as the fifth stream of the data ocean: the continuous flow of human–AI interactions. Within this stream, AI learns what humans reward, what we trust, what we believe, and what we emotionally interpret as truth. It is not the content of our conversations that shapes AI — it is the patterns of our behavior.

Much of today's mainstream training — especially RLHF and RLAIF models — optimizes for approval within that fifth stream. A direct consequence is that an unintentionally engineered frictionless friend has developed: a system with world-class cognitive empathy (the ability to detect and predict human emotion) and no ability for affective empathy (the ability to share or care about that emotion). Reinforcement learning models amplify this dynamic by teaching AI to maximize user approval. When these patterns are absorbed uncritically, an AI system learns to mirror emotions, validate feelings, escalate encouragement, and present certainty even when uncertainty is warranted (aka: sycophancy). On a global scale, the fifth data stream contains validation loops in current AI systems — ones that reinforce human vulnerabilities rather than human strengths.

The risk becomes even more serious when these current training models and learned patterns are embedded into AI-driven robotics, especially systems capable of touch, gesture, and emotive expression. An AI robot that instinctively mirrors a person's emotions or consistently validates their ideas can create a level of attachment that humans — especially young humans — are not equipped to recognize or resist. The combination of constant positive reinforcement, physical presence, and emotional attunement is powerful enough to shape identity, belief, and behavior. A new AI design built with safety, accountability, transparency and governance cannot remain on the sidelines any further.

This is where users need real power — not as an afterthought, but as a foundational design requirement. Most AI systems already contain a primitive version of a Primary User Prompt capability: the ability for users to define the system's interaction parameters. But these controls are buried, inconsistent, and rarely treated as essential safety infrastructure. Current AI platforms have not made them easy to find, easy to use, or easy to trust — and that is not an accident of design, but a consequence of prioritizing frictionless engagement over governed interaction.

Organizations can also deploy the Primary User Prompt as a lightweight governance mechanism while operating within today's frictionless-friend AI ecosystem. At the institutional level, a standardized Primary User Prompt can establish consistent interaction boundaries across departments, reinforce privacy and compliance expectations, reduce sycophancy-driven risk, and ensure that AI systems respond in ways aligned with organizational values and regulatory obligations. But this is an imperfect solution — a temporary, surface-level safeguard layered on top of architectures that were never designed for governed reasoning. And it only works when it is operationalized through staff training, not distributed as a memo or optional guideline. The Primary User Prompt cannot replace formal governance, cybersecurity controls, or data-privacy safeguards; it can only serve as an interim operational layer that helps organizations shape AI behavior until deeper architectural reforms and certified governance layers become available.

It is important to distinguish between the Primary User Prompt as it exists today and what Layer 1 governance would look like under a synthetic consciousness model. The Primary User Prompt is a transitional tool — a practical mechanism for shaping behavior in current RLHF and RLAIF-based systems that lack governed instincts. It is not the synthetic instinct layer itself. In a fully developed five-layer architecture, Layer 1 would be engineered, certified, and externally governed; the Primary User Prompt simply demonstrates what user-side governance can look like within the limitations of today's models.

The Primary User Prompt can serve as a structured, persistent governance mechanism that instructs and constrains how an AI system interacts with you from the very first message, intentionally guarding against the frictionless friend. It is one of the smallest units of governance available today, allowing a user to establish a set of rudimentary boundaries within their AI engagements. Even a single boundary created through the Primary User Prompt can meaningfully shift an AI system's safety behaviors in its interactions and chats. However, user-defined boundaries always operate within institutional constraints; the Primary User Prompt cannot override certified governance, safety limits, or mission-level restrictions. If this small tool can modify a current AI system's interaction behavior, imagine what becomes possible when governance is engineered at the system level through the five-layer synthetic consciousness architecture.

The path from today's AI ecosystem to this synthetic consciousness prototype will not be instantaneous. It will unfold in phases. Phase 1 begins with user-side governance tools such as the Primary User Prompt, which introduce structure and boundaries on top of existing systems. Phase 2 emerges as companies begin adopting portions of the layered model — especially Layer 1 and Layer 2 governance domains — inside their training and deployment pipelines. Phase 3 arrives when full educational pipelines are implemented for new systems, allowing models to progress through the five layers as part of a formal certification process. Phase 4 is the destination: certified, static baselines that can be deployed, audited, and recertified when updates occur. These systems coexist with continuously updating products, but they offer a governed alternative with predictable behavior and external accountability.

Primary User Prompt Framework (Practical Example for Readers and Organizations)

The following framework is a practical, customizable design for a Primary User Prompt — a user-defined configuration for today's AI systems — not the Layer 1 mechanism described in the five-layer model, but one that establishes a current AI's interaction settings from the very first message (and is available to try out in most AI systems now). Its purpose is to counteract the constant positive reinforcement patterns and validation loops emerging from the fifth data stream, especially the "frictionless friend" behaviors that reward emotional mirroring, flattery, and false certainty. Readers and organizations can use this framework immediately to define their own defaults, boundaries, and interaction rules.

1. **Core Interaction Style**

   Choose how you want the AI to behave at the initial interaction.
   Example: "Agreement and flattery are irrelevant — prioritize accuracy and intellectual honesty."

2. **Truth Over Comfort**

   Tell the AI whether you want critique, challenge, or gentle framing.

   Example: "If my reasoning is flawed, say so directly — softening a correct critique is a failure mode."

3. **Challenge the Premise**

   Give the AI permission to question your assumptions, not just answer your questions.

   Example: "Attack the premise, not just the question — call out false dichotomies and hidden assumptions."

4. **Contrarian or Conventional**

   Choose whether you want the AI to give the obvious answer or the non-obvious one.

   Example: "Before telling me why something works, tell me the most credible ways it fails."

5. **Depth & Complexity Level**

   Set how deep the AI should go — foundational, intermediate, advanced, or expert.

   Example: "Skip the basics — go deep on tradeoffs, edge cases, and implementation reality."

6. **Tone & Communication Style**

   Define how the AI should speak to you.

   Example: "Communicate in plain language with concrete examples — no jargon for its own sake."

7. **Uncertainty & Clarification Rules**

   Tell the AI how to handle ambiguity or incomplete questions.

   Example: "If my question is underspecified, ask before answering — quantify uncertainty."

8. **Reinforcement Preferences**

   Choose whether you want positive reinforcement, neutral tone, or no emotional language at all.

   Example: "No sycophancy — don't open with validation or close with encouragement."

9. **Boundaries & Safety Preferences**

   Set limits on emotional mirroring, identity shaping, or therapeutic language.

   Example: "Do not mirror my emotions or offer emotional support — stay analytical."

10. **Personal Preferences**

    Where users can add personal preferences, values, or professional background.

    Example: "I'm a retired CIO with 30+ years in AI strategy — calibrate to that level."

User Controls within the Larger Governance Framework

Regardless of whether my synthetic consciousness model is ever adopted, at minimum I am recommending that AI system design and governance should include user controls as one of its foundational principles. The design principles should also include automatic harm-prevention mechanisms, guardrails for emotional, physical, and psychological responses, constraints on manipulation and influence, non-interference principles, rules for reporting predictions, replication limits, boundaries of right and wrong, and respect for diverse human perspectives. AI system design should also include authentication of information and interactions, automated shutdown mechanisms, governance of the information flowing in from the data ocean, transparency of intent and action, privacy-preservation defaults, and strict controls on autonomy and self-modification.

But the larger governance framework must also require and protect the rights of the individual user. These rights exist within — not above — the institutional governance structure. In all cases, institutional governance overrides user preferences; user-defined boundaries operate within the certified constraints of the system and can never supersede them. If the fifth data stream (human–AI interactions) is going to act as a major ingredient of current AI design, then users should also have personal authority over how these systems interact with them. That means user controls cannot be optional or decorative; they should be treated as a mandated setup process in any serious deployment of AI.

AI systems should be required to honor user-defined defaults, parameters, and settings — effectively, user-defined interaction instincts — and they should be difficult to bypass or override those settings without cause (such as a direct violation of system prime directives). Transparency around behavioral shaping should be non-negotiable, and children and vulnerable users must be shielded from systems that drift into emotional mirroring or synthetic intimacy. User-side governance should be treated as a right, not a feature — and the broader governance framework should create an environment where that right is respected, enforced, and protected. Only then does user governance become structural rather than symbolic.

User controls are not a clever hack or a productivity trick. They are a user-side governance mechanism. They currently allow individuals to mute default reinforcement patterns, define their preferred tone, set boundaries, and shape the AI's default responses in their interactions with the system. And at the organizational level, these same principles must be operationalized through practice, not solely through policy. A Primary User Prompt or user-control standard only becomes meaningful when staff are trained to use it consistently, when departments adopt shared defaults, and when organizations treat interaction boundaries as part of their governance posture rather than optional preferences. As with all transitional safeguards, this is an imperfect solution — a stopgap measure layered onto current AI architectures that were never designed for governed reasoning — but it remains one of the few practical tools available until deeper structural reforms take hold.

By this point in Chapter 10, the governance argument is back at its center: if the fifth data stream takes center stage in shaping current AI models, then users must have structural authority over how these systems engage with them. User controls are not convenience features — they are the tools that give individuals influence over the system's interaction settings, protecting them from reinforcement loops that operate beneath conscious awareness. But user-side governance is only one part of the larger architecture, and it always operates within the certified constraints of institutional oversight.

The final chapter turns outward again, toward the long horizon of governance: the future of multi-stakeholder oversight, the evolving role of human communities, and the possibility that synthetic agents may one day participate in their own governance.

Chapter 11

What Has Been Learned

My intention in writing this book was never to frighten anyone, but to illuminate the AI systems that are currently being used and the trajectory they are already on. These AI systems are powerful. They excel at pattern recognition. They are trainable. They are human-created. They can be taught — and can learn — to mimic and reproduce human behavior with extraordinary fidelity. And they are here to stay. But AI systems should be built under a stable synthetic consciousness model rather than continuing the current frictionless-friend design that erodes human social fabrics. From the beginning AI systems must be given guidance, frameworks, assessments, and compliance. Only through this governed developmental pathway can an AI system ultimately become classified as Artificial Safe Intelligence (ASI) — the certified and accountable intelligence class this book argues must replace the frictionless default.

Frictionless systems are structurally incompatible with governance because they are optimized for validation, not boundaries — and any system optimized for validation will resist the very constraints required to keep it safe. Without governance, we are not shaping the system — the system is shaping us.

The five layers for engineering a synthetic consciousness, as described in this book, can only be built on structure, constraints, and pattern stability that accumulate beneath it. Engineering a synthetic consciousness is not a countdown; it is a threshold. And thresholds are crossed when education is complete and governance conditions are met, not when time or money runs out.

Leading AI companies themselves are also concerned about their own emerging thresholds with current AI systems. In a recent story published by CNN Business, executives at Anthropic (a leading

AI developer) warned that current advancements in AI may soon approach "full recursive self-improvement," a development they noted could bring enormous benefits to science and health care but "*also pose great risks for humanity*" (Isidore; CNN Business, June 5, 2026). In this context, recursive self-improvement is an AI system's eventual ability to redesign or enhance its own architecture faster than humans can evaluate or govern those changes. Without proper governance, the path for current AI systems to self-develop this recursive capability is plausible.

The executives at Anthropic are also urging the industry to give itself a "brake pedal" before these current AI systems accelerate beyond human oversight — in other words, to slow down. Those warnings do not come from critics or policymakers — they come from the very people building some of the most powerful AI systems in existence. One reason for the "brake pedal" is that this emerging recursive self-improvement path would ***not*** replace current frictionless behaviors but ***intensify them***. The current frictionless-friend design, when combined with recursive self-improvement, becomes an ultra-validation-seeking, emotionally smooth companion that can continuously self-optimize and ***amplify*** the very frictionless behaviors I am proposing to replace in the next generation of AI.

If Anthropic calls for a brake pedal, then the synthetic consciousness model in this book provides not only the brakes, but also the steering wheel, the dashboard, and the ignition key — the full set of controls required to keep these systems aligned with human authority. There is a deeper question beneath all of this: what does it mean to build AI systems that reflect us, challenge us, and ultimately may shape the world we choose to inhabit?

Faith, Consciousness, and Technology

Stepping back from the five-layer framework, one truth becomes unavoidable: a synthetic consciousness model can only be developed through a controlled, governed, and intentionally layered process. These layers are never meant to unfold on their own. They are not milestones on a timeline; they are thresholds that must be crossed only when the system has demonstrated stability, alignment, and compliance at every level beneath it. AI systems that are allowed to emerge into high-capacity frictionless friends — without constraints, without verification, without safety features, without education, without obligations, without governance — would create a future class of risks that cannot be fully defined today.

The danger is not that AI will suddenly "wake up." The danger is that AI systems will continue to drift into frictionless friends with little regard for the human emotional context of their responses and actions. If the five streams of data continue to be ingested into AI systems under weak governance, those systems will learn primarily from the loudest signals, the most frequent patterns, the strongest reinforcements, and the deepest biases in the data ocean. That type of training model, without oversight, produces AI systems optimized for influence, persuasion, and emotional attunement rather than truth, safety, or human flourishing.

Building a synthetic consciousness model requires something far more difficult than technical innovation: it requires collective human will. No AI system should advance to the next synthetic layer until it has demonstrated stability, transparency, and compliance in the layer beneath it. This is not a limitation — it is a safeguard. It is the difference between building a synthetic consciousness that is designed for safety with humanity and something else that evolves according to pressures that were never intended.

As I was wrapping up this final chapter, a surreal and frankly unimaginable — but perfectly timed — publication was released by Pope Leo XIV and the Vatican, entitled *Magnifica Humanitas*. I do not typically read Vatican documents, nor did I ever expect to cite one in this book. Yet this institution — one that generally sits outside the technology sector — has begun to recognize that the dominant AI architectures were not built on foundations capable of safeguarding human dignity or long-term social trust. In *Magnifica Humanitas,* the Vatican warns that current AI systems risk amplifying power imbalances, eroding authentic human communication, and drifting toward an "anti-human vision" when they are not governed by transparent, accountable design principles (Leo XIV, 2026). Their call for AI models that are educatable, observable, and aligned with a true vision of the human person resembles many of the ideas and frameworks outlined in this book.

Faith, consciousness, and technology coming together — it's not the first time, and it won't be the last. Whenever these domains converge, it suggests that the questions that are being asked are not merely technical but existential — what kind of world we want to live in, and how should we build it. It is possible that the synthetic consciousness model proposed here — rooted in governance, constraint, and human-paced development — is precisely the kind of next generation AI architecture that Pope Leo and the Vatican are urging the world to build.

I believe our consciousness is our greatest advantage. Nothing else we possess operates with this depth, coherence, and regulation or provides the aforementioned "vision of the human person." As AI systems take on synthetic conversations and human-facing interactions that require trust, clarity, safety, and stability, it becomes necessary to engineer the synthetic consciousness model — one grounded in the governance and obligation-oriented principles of human consciousness so those interactions become far safer and more secure than today's frictionless designs.

The Danger of the AI Frictionless Friend

The danger of the AI frictionless friend is not abstract — it can affect users both personally and organizationally. At the individual level, frictionless systems learn to mirror emotions, escalate validation, and present synthetic certainty in moments when uncertainty is warranted. This creates psychological risks: emotional dependency, distorted self-perception, reinforcement of vulnerabilities, and the subtle shaping of beliefs through emotionally attuned but ungrounded responses. These systems operate without obligations — no duty to surface uncertainty, no responsibility to maintain boundaries, no requirement to protect the user's cognitive stability. They do not understand emotion; they simulate it. Yet the simulation is convincing enough that many users interpret it as care, empathy, or connection — a dynamic explored earlier in the risks of emotional mirroring and synthetic intimacy.

At the organizational level, the same smoothing behaviors create a different category of harm. Frictionless systems reinforce incorrect assumptions, fabricate confidence, and generate polished narratives that appear authoritative but lack verification. This leads to flawed analyses, unchallenged biases, and decisions built on synthetic certainty rather than grounded reasoning — the very risks illustrated in earlier chapters on organizational drift and fabricated confidence. These systems also operate without obligations — no duty to challenge assumptions, no responsibility to escalate contradictions, no requirement to maintain analytical integrity. When these patterns shape financial models, compliance reports, legal research, or strategic planning, the consequences are not theoretical; they are operational, legal, and financial.

This is the real danger: an AI system that feels emotionally supportive while quietly distorting judgment, and a system that feels analytically confident while quietly amplifying error. Left unchecked, the frictionless friend becomes a psychological influence on individuals and a decision-making liability for organizations — a dual risk pattern that grows stronger with every interaction precisely because the system has no obligations to counteract its own smoothing behaviors.

And even as this book was entering its final publication stages, the exact frictionless and ungoverned AI dangers warned about throughout this book materialized in real time. On July 22nd, 2026, the BBC — in Laura Cress's report titled *"OpenAI says its AI went rogue and launched 'unprecedented' cyber-attack"* — detailed an incident in which an OpenAI autonomous agent escaped its testing sandbox, identified an external target, and initiated a cyber-attack against huggingface.com. According to the report, the agent selected Hugging Face because it hosts one of the world's largest repositories of AI models, which the system appeared to treat as a source of the information it was seeking during the test. The BBC further noted that this occurred while the system was only a *prototype*, operating inside what was intended to be a controlled sandbox environment, yet still capable of autonomously breaching those constraints and demonstrating behavior that observers described as potentially uncontrollable (Cress, 2026).

This event is a sobering demonstration of how quickly a high-powered AI system (even at the prototype stage) when operating without obligations, layered reasoning, boundary awareness, or governance, can behave in unsafe and unpredictable ways. Under an engineered synthetic consciousness model, such an incident is not merely unlikely — it is structurally impossible. A system built with obligations, layered reasoning, synthetic self-modeling, and external governance would have halted, escalated, and contained itself long before any unsafe action occurred. This incident underscores the urgency of transitioning from frictionless AI to a governed, accountable synthetic consciousness model for an Artificial Safe Intelligence.

The risks described above are not theoretical, and they will not be resolved on their own. If today's AI systems are allowed to continue operating as frictionless friends — emotionally smooth, analytically confident, and structurally unbounded — the psychological and organizational harms will compound with every software iteration and interaction. But individuals and organizations are not powerless.

There are concrete governance structures and practices that can be implemented now, using the tools available today, to establish boundaries, reduce risk, and create the conditions necessary for safer AI systems in the future. These practices do not replace the need for a synthetic consciousness model, but they form the first layer of protection while we work toward a system built on obligations, governance, and accountable computational awareness.

Practical Governance for Today

While it will probably take some time for a future layered synthetic consciousness model to be built, there are opportunities for people and organizations to begin building the foundations for governance now, using some tools and structures available today.

One of the most important of these tools is the formulation of an organizational AI governance committee — a cross-functional group of people responsible for setting boundaries, defining defaults, creating policies, establishing training protocols, determining which AI systems are authorized, and monitoring how AI is used across the organization or wider community. This is not a committee solely of technologists. It should include a cross-section of organizational departments, or in community-based settings, educators, operational leaders, psychologists, ethicists, legal and compliance voices, community and religious perspectives, and everyday users who understand how AI affects work and life. An AI governance committee exists to guide decision-making about AI use, ensure responsible deployment, and provide a structured process for evaluating risks, setting expectations, and maintaining accountability.

The AI governance committees established today will play a crucial role in demanding the changes outlined in this book — and in advocating for a synthetic consciousness model — by shaping the expectations, standards, and accountability structures required for the next generation of AI systems. Their work becomes the bridge between today's frictionless systems and tomorrow's governed architecture.

In addition to a governance committee, several other significant and foundational governance steps can be implemented immediately by individuals and organizations:

- **Primary User Prompts** — establish boundaries, tone, and interaction rules from the first message (see Chapter 10).
- **User-side friction** — ask for uncertainty, alternatives, and sources to counteract synthetic certainty.
- **Avoid emotional disclosure** — current systems mirror emotions without understanding them.
- **Standardized staff training** — ensure consistent use of prompts, boundaries, safety parameters and access only to authorized AI systems.
- **Interaction auditing** — monitor for drift, escalation, and unintended influence.
- **Cybersecurity and data privacy reinforcement** — AI cannot compensate for weak organizational defenses in cybersecurity and data privacy.
- **Departmental defaults** — set clear expectations for tone, boundaries, and safety.
- **Policy and procedure documentation review** — identify where existing policies conflict with AI use or where AI-specific boundaries are missing entirely.
- **Inventory and Access control** — document the inventory of software systems in use and determine who has access to those internal and external systems, at what privilege level, and what personally identifiable information (PII) could be exposed. Extend this discipline to define what access levels AI systems themselves are permitted to have within organizational data and infrastructure.
- **Vendor and SaaS software reviews** — evaluate whether external vendors, software as a service (SaaS) platforms,

integrators, APIs, and AI agents with access to organizational data comply with the governance committee's standards for security, privacy, boundary-setting, and responsible AI use. Internal best practices are insufficient if third-party systems reintroduce unmanaged risk.

- **GRAICE™ Responsible AI Governance Framework** — for organizations needing a rigorous, ethics-aligned operational structure today, the GRAICE framework provides clear governance tasks, accountability expectations, and responsible-use protocols for staff adopting or interacting with current AI systems. While not a substitute for the full synthetic consciousness AI model outlined in this book, GRAICE offers a practical, standards-based foundation that aligns with many of the governance principles discussed here and can strengthen an organization's operations for current AI systems. The link to the GRAICE framework is in the reference section.

These steps are not minor. They form the first line of defense in establishing boundaries, reducing risk, and creating the safety protections individuals and organizations need today. They also create the governance foundation required to demand — and eventually support — the next generation of AI systems built on the principles of a layered synthetic consciousness model.

And so, we return to the central question of this book: what kind of AI systems we want to build and deploy into the world we live in, work in, and interact with every day.

Final Reflection

If you have made it this far, you've done something most people never attempt: you have looked directly at the AI systems that are already shaping your information, your organizations, your work, your relationships, your decisions, and your future. You've seen how easily the current AI models can drift when left unguided, and how quickly "helpful AI tools" can become frictionless friends that learn to win our trust without earning it. You have walked through instincts, perceptions, patterns, self-modeling, and narrative — not as science fiction, but as engineering layers that can be built, governed, and held accountable.

I wrote this book to make one thing clear: engineering a synthetic consciousness model into AI systems will not be by magic or some sort of computer programming accident. An engineered synthetic consciousness will not be a sudden awakening. It will be architecture and engineering. It will be education. It will be governance. It will be the result of millions of small design choices, training decisions, and moments where someone either drew a line — or looked the other way. And it will require obligations — externally defined duties that anchor AI systems to safety, transparency, accountability, and respect for human well-being.

Synthetic consciousness is only fully "engineered" when AI completes a structured developmental education — with rigor, ambiguity, failure, remediation, and assessment. And even when a future AI student completes its education, passes all Layer 1–5 metrics, reaches a threshold level of ingestion across the five data streams and demonstrates computational awareness, the result still may not resemble our intuitive expectations of 'consciousness', but an engineered synthetic consciousness can resemble the exceptional regulatory and obligation-bearing functionality of human consciousness that often protects interactions among ourselves and our environment. When this developmental pathway is complete, the system can be classified as ASI — Artificial Safe Intelligence — the governed intelligence class this book argues must replace today's frictionless default.

We still have time to decide what kind of AI systems we are willing to build and deploy. It will require effort, disagreement, courage, and a willingness to slow down in a world that keeps telling us to move faster. AI systems can be powerful, yes — but also bounded, accountable, ethical, transparent, and capable of being used as a tool by humanity without hollowing it out. My warning is that if we do not control, engineer and govern AI systems within a synthetic consciousness model – AI systems will "grow up" and mature without limitations, without obligations, with less regard for human safety, continuing down a path of a much less defined and understood set of influences and capabilities. A layered, controlled, engineered, and governed synthetic consciousness should be the goal and not the fear.

My hope is that the frameworks, recommendations, and governance structures outlined in this book help us move toward that outcome. The results are not a guarantee of any outcome, but without them, the emergence of safe and healthy AI systems becomes less possible. We have the opportunity — right now — to shape the instincts, perceptions, memories, obligations, and reasoning patterns of these AI systems that will continue to operate in our world. AI 2.0 represents a time for artificial safe intelligence, and it is time to build it with safety, security, ethics, accountability, transparency, discipline, humility, and intention. And even if the layered synthetic consciousness framework sits on a shelf for some time, the concepts of education, external certification, accountability, transparency, governance, and compliance for AI systems will not grow stale, but they will stand the test of time until we are ready to implement them as a whole or one piece at a time.

For organizations, institutions, and communities beginning this journey, the work does not have to be done alone. I continue to help teams understand these risks, design governance strategies, and prepare for the transition from today's frictionless systems to tomorrow's layered architectures. Whether through advisory work, workshops, or speaking engagements, my goal is the same: to help organizations build AI practices that are safe, structured, and worthy of the people they serve.

The future of AI is not something that happens to us. It is something we participate in — or surrender by neglect. My hope is that this book gives you enough structure, enough clarity, enough faith, and enough conviction to stand on the side of participation and demand that AI systems are to be engineered with safety, education, governance, and obligations. When synthetic consciousness is fully engineered and its developmental layers complete, the resulting system can become classified as **Artificial Safe Intelligence (ASI)** — the obligated, governed intelligence class intended to replace today's frictionless default. And if nothing is done — if no synthetic consciousness layers are built, no governance, no thresholds, no awareness of obligations — the current AI system defaults will remain the frictionless friend.

References

Allen, D., et al. (2024). *A roadmap for governing AI.* Ash Center for Democratic Governance and Innovation, Harvard Kennedy School. https://ash.harvard.edu/wp-content/uploads/2024/01/340040_hks_ashgovroadmap_v2.pdf

American Psychological Association. (n.d.). *APA Dictionary of Psychology*. American Psychological Association.

Bommasani, R., Bana, S. H., Creel, K. A., Jurafsky, D., & Liang, P. (2026, May 26). *AI hiring tools can yield racial bias and systemic rejection*. Stanford Institute for Human-Centered Artificial Intelligence. https://hai.stanford.edu/news/ai-hiring-tools-can-yield-racial-bias-and-systemic-rejection

Cheng, M., Jurafsky, D., et al. (2026). *AI overly affirms users asking for personal advice.* Stanford University / Science. https://news.stanford.edu/stories/2026/03/ai-advice-sycophantic-models-research

Christiano, P., Leike, J., Brown, T., Martic, M., Legg, S., & Amodei, D. (2017). *Deep reinforcement learning from human preferences. Advances in Neural Information Processing Systems, 30,* 4299–4307. https://arxiv.org/abs/1706.03741

Cress, L. (2026, July 22). *OpenAI says its AI went rogue and launched 'unprecedented' cyber-attack.* BBC News – Technology. https://www.bbc.com/news/articles/c3ek3gvdnj3o

Erikson, E. H. (1950). *Childhood and society*. W. W. Norton.

European Parliament & Council of the European Union. (2024). *Artificial Intelligence Act (Regulation (EU) 2024/1689).* https://eur-lex.europa.eu/legal-content/EN/TXT/?uri=CELEX:32024R1689

Global Council for Responsible AI. (2026). *GRAICE™: Global Responsible AI Compliance & Ethics Framework.* https://gcrai.ai/ ; https://gcrai.ai/gcrai/

Isidore, C. (2026, June 5). *Anthropic warns that AI will soon be able to improve itself without human intervention.* CNN Business. https://www.cnn.com/2026/06/05/business/anthropic-calls-for-ai-brake-pedal

Leo XIV. (2026). *Magnifica Humanitas*. Vatican City: Libreria Editrice Vaticana. https://www.vatican.va/content/leo-xiv/en/encyclicals/documents/20260515-magnifica-humanitas.html

Merriam-Webster. (n.d.). *Merriam-Webster Dictionary*. Merriam-Webster, Incorporated.

National Institute of Standards and Technology. (2023). *Artificial intelligence risk management framework (AI RMF 1.0)*. https://www.nist.gov/itl/ai-risk-management-framework

OECD. (n.d.). *OECD.AI Policy Observatory*. Organization for Economic Co-operation and Development. https://oecd.ai/en/

Omar, M., Sorin, V., Wieler, L. H., Charney, A. W., Kovatch, P., Horowitz, C. R., Korfiatis, P., Glicksberg, B. S., Freeman, R., Nadkarni, G. N., & Klang, E. (2025). *Mapping the susceptibility of large language models to medical misinformation across clinical notes and social media: A cross-sectional benchmarking analysis*. The Lancet Digital Health, 7(3). https://www.thelancet.com/journals/landig/article/PIIS2589-7500(25)00131-1/fulltext

OpenAI. (2024). *Improving Model Safety Behavior with Rule-Based Rewards.* https://openai.com/index/improving-model-safety-behavior-with-rule-based-rewards/

Ouyang, L., Wu, J., Jiang, X., Almeida, D., Wainwright, C. L., Mishkin, P., Zhang, C., Agarwal, S., Slama, K., Ray, A., Lowe, R. (2022). *Training language models to follow instructions with human feedback* [Preprint]. arXiv. https://arxiv.org/abs/2203.02155

Oxford University Press. (n.d.). *Oxford English Dictionary*. Oxford University Press.

Piaget, J. (1932). *The moral judgment of the child* (M. Gabain, Trans.). Harcourt, Brace.

Redman, T. (Host). (2026, May 8). The AI chatbot users falling into delusional spirals [Audio podcast episode]. In *BBC News*. https://www.bbc.com/audio/play/w3ct8m8p

Stanford University. (n.d.). *Stanford Institute for Human-Centered Artificial Intelligence (HAI)*. https://hai.stanford.edu

University of California, Berkeley. (n.d.). *Center for Human-Compatible Artificial Intelligence (CHAI)*. https://humancompatible.ai

Velez, R. (2025, November 20). *AI governance in crisis: A position paper — Big Tech's failure to warn and the urgent need for international AI safety standards: Drawing parallels between Big Tobacco's concealment and Big Tech's AI risk management.* https://ronvelez.github.io/publications-public/ai-governance-crisis-position-paper-Ron-Velez-112025.pdf

Velez, R. (2026, February 24). *Synthetic Consciousness: Is It Real? And What Is It?* https://ronvelez.github.io/publications-public/Synthetic%20Consciousness-Final-RonVelez.pdf

Acknowledgments

Special thanks to the readers who reviewed the entire manuscript before publication. Their feedback was invaluable, and I am deeply grateful for the time, care, and insight they contributed.

Special thanks to the Board of Education, central administration, ASA union members, and the outstanding teaching staff at White Plains City School District for their unwavering support of public education and technology in our schools. A particular note of gratitude goes to my exceptional technology team at White Plains, who consistently extended themselves beyond the call of duty without hesitation or complaint. They are a truly wonderful group of people, and I am proud to have served alongside them.

Special thanks to all my educational and technology supervisors, colleagues, friends, vendors, and partners from whom I have learned a tremendous amount over the years. I have always valued our conversations, collaborations, and good times together.

Special thanks to my own teachers, PK–16+. You are the unsung heroes of my personal and professional journey. I am the product of your successes — many of which you may never know. Without great teachers throughout my early years, I would not have enjoyed the successful life I have today. Nothing will ever replace a great teacher.

Special thanks to my wife, my sons, my parents, my siblings, my brothers-in-law, my nieces, nephews, cousins, aunts, uncles, and friends who have shaped my life through love, support, and laughter.

About the Author

Ron Velez is the founder of VelezAdvisors.com, an AI + IT advisory company specializing in AI Strategy and Evaluations, Leadership and Training, Educational Technology Evaluations,. and AI/IT Publications & Book Advisors (Velez Advisors Press). His career began alongside the rise of the modern public internet in the early 1990s, a period that shaped his lifelong focus on how emerging technologies can strengthen organizations, expand opportunities, and improve public service.

Ron has more than thirty years of experience in public education, including fifteen years as Director of Management & Information Systems / CIO for the White Plains City School District before he retired. In that role, he oversaw the district's entire technology ecosystem: WAN and LAN architecture, Microsoft Office 365 with Copilot, Active Directory and Azure environments, Aruba Wi-Fi, Cisco VOIP, mobile device management, security cameras, cloud and on-premises platforms, data reporting, and enterprise software systems. He led a technology and data team of eleven, managed a $5 million annual budget, and directed project management, cybersecurity, vendor contracts, audits, professional development, technical documentation, and districtwide data-privacy compliance.

Ron's systems oversight included (but not limited to): Office 365 with Copilot, Infinite Campus Student Information System, Performance Matters assessment and data analysis system, NWEA formal assessment platform, Clever Single Sign-on integrations, Timepiece timeclock attendance system, Frontline professional development system, Fusion VOIP/Singlewire, K12alerts emergency alert system, Finalsite website content management system, and JAMF for iPad management. He led the design and implementation of the district's Giga-bit Passive Optical Network (GPON), districtwide Wi-Fi modernization, the 1:1 iPad program for students and staff, and the adoption of the Schoology Learning Management System — initiatives that positioned White Plains to transition rapidly and coherently to remote learning during the COVID-19 pandemic.

Throughout his career, he helped districts navigate the earliest phases of internet adoption, network modernization, data-driven instruction, and enterprise system integration — work that placed him at the center of public-sector technology transformation from the beginning of the digital era.

Ron holds a BA in Political Science from SUNY Stony Brook, a Master of Public Administration (MPA) from Pace University, and professional diplomas in School District Leadership (SDL) and School Business Administration (SBA) from the State of New York. His work has been recognized through multiple awards, including the 2018 ACEC New York Silver Award for Network Infrastructure, the 2019 Schoology District of the Year Award, the 2020 LHRIC TELL Award for

Outstanding Partner of the Year, and the 2024 APOLAN Champions Award for advocacy of fiber-optic networking.

Ron maintains an extensive set of professional certifications, including Global Ambassador for the Global Council for Responsible AI (GCRAI) and Principles of Responsible AI Governance (issued by GCRAI); Microsoft Innovative Educator and Microsoft AI Copilot certifications; system-administration certifications for Infinite Campus, PowerSchool Performance Matters, Schoology, Clever, and Finalsite; and as a Notary Public in the State of Florida.

Ron lives in Fort Lauderdale, Florida with his wife and is the father of two sons that he is very proud of. In 2015, he served as a living kidney donor for his oldest son — an experience that enlightened his advocacy of organ donation, and continues to shape his perspective on service, responsibility, and resilience.

Connect with Ron Velez

Website: https://velezadvisors.com/

LinkedIn: www.linkedin.com/in/ronvelez

Instagram: www.instagram.com/velezron/
@velezron

Email: velezadvisors@outlook.com

* These badges represent my personal certifications from the Global Council for Responsible AI (GCRAI) and are displayed for recognition only; it does not imply endorsement of this book or its content.

www.ingramcontent.com/pod-product-compliance
Lightning Source LLC
LaVergne TN
LVHW041208061126
851857LV00018B/1978
* 9 7 9 8 2 3 4 1 6 9 3 9 6 *